Successful Management of the Analytical Laboratory

Oscar I. Milner

CRC Press
Taylor & Francis Group
Boca Raton London New York

CRC Press is an imprint of the
Taylor & Francis Group, an **informa** business

First published 1991 by Lewis Publishers, Inc.

Published 2019 by CRC Press
Taylor & Francis Group
6000 Broken Sound Parkway NW, Suite 300
Boca Raton, FL 33487-2742

© 1991 by Taylor & Francis Group, LLC
CRC Press is an imprint of Taylor & Francis Group, an Informa business

First issued in paperback 2019

No claim to original U.S. Government works

ISBN-13: 978-0-367-45044-1 (pbk)
ISBN-13: 978-0-87371-438-9 (hbk)

This book contains information obtained from authentic and highly regarded sources. Reasonable efforts have been made to publish reliable data and information, but the author and publisher cannot assume responsibility for the validity of all materials or the consequences of their use. The authors and publishers have attempted to trace the copyright holders of all material reproduced in this publication and apologize to copyright holders if permission to publish in this form has not been obtained. If any copyright material has not been acknowledged please write and let us know so we may rectify in any future reprint.

Except as permitted under U.S. Copyright Law, no part of this book may be reprinted, reproduced, transmitted, or utilized in any form by any electronic, mechanical, or other means, now known or hereafter invented, including photocopying, microfilming, and recording, or in any information storage or retrieval system, without written permission from the publishers.

For permission to photocopy or use material electronically from this work, please access www.copyright.com (http://www.copyright.com/) or contact the Copyright Clearance Center, Inc. (CCC), 222 Rosewood Drive, Danvers, MA 01923, 978-750-8400. CCC is a not-for-profit organiza-tion that provides licenses and registration for a variety of users. For organizations that have been granted a photocopy license by the CCC, a separate system of payment has been arranged.

Trademark Notice: Product or corporate names may be trademarks or registered trademarks, and are used only for identification and explanation without intent to infringe.

Visit the Taylor & Francis Web site at
http://www.taylorandfrancis.com

and the CRC Press Web site at
http://www.crcpress.com

Library of Congress Card Number 91-30754

Library of Congress Cataloging-in-Publication Data

Milner, Oscar I. (Oscar Irving), 1914-
    Successful management of the analytical laboratory / Oscar
I. Milner.
    p. cm.
Includes bibliographical references and index.
ISBN 0-87371-438-5
    Chemical laboratories—Management. 2. Chemistry, Analytic.
I. Title.
QD51.M62 1991
542'.068—dc20   91-30754

Preface

The position of the analytical laboratory, particularly in an industrial organization, is often a difficult one. The laboratory has traditionally been viewed as a service group whose main function is to provide information on composition for others to use in making decisions. Consequently, the laboratory is not only subject to competing pressures for its services, but its staff has sometimes been looked upon as being professionally less qualified than scientists and technologists in other areas.

This book reviews the operation of the analytical laboratory from the standpoint of the manager's responsibilities. It offers a number of suggestions to help improve the level of the laboratory's performance and its cost-effectiveness as well as the professional status of its personnel. The material is based on a seminar that the author has conducted for laboratory managers since 1980. Although the presentation is intended for managers at various levels of experience, it should be particularly helpful to relatively new and potential managers by defining problems they are likely to encounter and offering possible solutions to them.

Because of the wide range of activities that use the services of analytical laboratories, the specific concerns of individual types of organizations are not addressed. The author has found, however, that most of the problems faced by analytical laboratory managers are remarkably similar, regardless of the laboratory's affiliation. Hence, although written from the perspective of one whose experience has been largely in an industrial environment, the book should prove useful to laboratory managers in other segments of our society as well.

The opening chapters deal with the functions and organization of typical laboratories. Succeeding chapters deal with managerial problems related to staffing, management and development of personnel, communication, employee safety, sample handling, workload control, quality performance, and financial and information management. Although managers do, of course, have ultimate responsibility for all decisions relating to the technology used in the laboratory, such technical matters as are discussed are limited to those that have a direct impact on their responsibilities in the above areas.

Acknowledgments

The author wishes to acknowledge the helpful suggestions of Dr. John K. Taylor, who reviewed sections of the chapter on quality performance, and those of Dr. Robert J. Kobrin, who reviewed the chapter on information management. Valuable contributions were also made by numerous laboratory managers who, during seminars conducted by the author, contributed valuable insights into the problems they face and means of dealing with them. Many of the concepts discussed in this book include ideas expressed by those managers, and their sharing of experiences on the "firing line" has contributed significantly to this presentation. It is a privilege to have been associated with so many dedicated colleagues.

The Author

Oscar I. Milner has a broad background in analytical laboratory management. He has spent over 40 years in the laboratory as an analytical chemist at various levels of supervision and management. His responsibilities have included such typical management duties as organization, facilities planning, project coordination and evaluation, staffing, personnel development, quality assurance, interlaboratory and academic relations, budgeting, and financial control. He has conducted numerous in-house management training courses and has often been called upon to participate as an outside expert in laboratory evaluations in the U.S. and abroad.

Mr. Milner received a Master's degree in chemistry from Temple University, was elected to Sigma Xi, and during World War II was a project leader and laboratory supervisor at the U.S. Navy's Industrial Test Facility at the Philadelphia Naval Base. He subsequently joined Mobil Corporation's Research and Development Laboratory, from which he is now retired. He is the author of 25 technical publications and the book *Analysis of Petroleum for Trace Elements,* and has served as co-editor of an American Chemical Society publication, *Analysis of Petroleum for Trace Metals.* Since leaving Mobil, he has acted as a laboratory management consultant and presented some 40 seminar sessions in the U.S. and Canada on management of the analytical laboratory.

Contents

Case Histories

INTRODUCTION

Role of the Analytical Laboratory in Today's World

Few branches of chemistry have shown as remarkable a rate of growth as analytical chemistry has during the past several decades. Much of this growth is attributable to the proliferation of environment-related problems and a large increase in the number of laboratories and personnel dealing with those problems. Greater concern with quality control is another factor in the increased emphasis on obtaining compositional information. Today, analytical laboratories are found in almost all segments of our society, including business and industry, agriculture, the military and various other branches of government, research institutes and universities, and health sciences. It has been estimated that these laboratories make 250 million individual measurements a day, at an annual cost of $50 billion.[1]

The greatest number of laboratories are undoubtedly those associated with industrial organizations, such as chemical, food, petroleum, pharmaceutical, transportation, mining, water and power. In these areas the analytical laboratory has always served an important function by providing data in support of other branches of science and engineering and in helping control product quality or process variables. In recent years, however, the laboratory has in many cases come into its own as a semi-independent entity, geared to the solution of problems by means of the techniques available to it, rather than serving only to provide data for others to interpret. Whether these problems are solved independently or by co-operative effort is not important. What is important is that the broad capabilities of the laboratory be recognized. Where this is the case, the laboratory can exercise a unique function in developing information essential to the organization of which it is a part, and its staff will enjoy the prestige among its peers that it merits.

REFERENCE

1. **Hertz, H.S.,** Are quality and productivity compatible in the analytical laboratory? *Anal. Chem.* 60(2): 75A–80A (1988).

1

Functions of the Industrial
Analytical Laboratory

The function of many of today's analytical laboratories is not always easily described. To those in the laboratory it may seem to be at the core of all operations, since the information it provides affects major decisions that can determine the very existence of the organization it serves. To others, it is merely a convenient way to obtain data on which to base their actions so that they can spend most of their time pursuing more productive activities. In fact, elements of truth are to be found in both concepts. The degree to which one or the other of these views predominates depends on the broad nature of the analytical work being performed.

Generally, the function of the industrial analytical laboratory is designated as process or product quality control, technical service, or research and development. Often, however, two or more of these functions are exercised in the same laboratory, sometimes by the designation of personnel for a specific type of assignment. It should be emphasized that although the majority of analytical laboratories may be those serving industry and can be placed in one of these four categories, many operations and problems are similar, regardless of the laboratory's affiliation, and can be viewed from the same perspective.

THE PROCESS CONTROL LABORATORY

The process control laboratory is responsible for providing the data that are used by process engineers and operators to

ensure that variables associated with the operation of a manufacturing, refining, or similar type of plant are kept within
prescribed limits. It is preferably located as close to the plant
as possible so as to minimize delay in the handling of samples
and to facilitate person-to-person communication. Often, a
number of different processes are to be monitored, and since
operational control may require results to be promptly available, sampling and testing are usually conducted on a scheduled frequency. The laboratory normally will also be responsible for analyzing raw materials and intermediates that enter
into the process(s) being controlled. If, as is often the case,
suppliers' certificates of analysis are accepted, it is the laboratory's responsibility to have ensured the reliability of such
data.

Since many plants operate continuously, shift work on the
part of the laboratory is often required. Tests are usually of a
standard nature and run on a repetitive basis. They are thus
normally conducted by technicians or low-level professionals
as an entry-level assignment. Occasionally, when a plant encounters an upset, the usual test procedures may be inadequate to identify the source of the problem. In such cases, the
services of more experienced professional chemists using nonstandard analytical methods may be required.

THE PRODUCT QUALITY CONTROL LABORATORY

The product quality control laboratory incorporates many of
the features of the process control laboratory. Tests may be
run on raw materials as well as on intermediates and final
products. Although the time element may sometimes be less
critical, there are occasions when immediate data are required,
as when controlling a blending operation or approving ship-
ment of a finished product.

The laboratory will also often become involved with suppliers and customers if questions arise with respect to composition or deviation from specification limits. Analytical procedures are generally fairly well standardized; they are also
likely to be repetitive in nature and carried out by personnel
at the same technical level as in the case of the process control
laboratory. Here, too, however, some upset condition may

make it necessary to use nonstandard methods and the services of more experienced professional chemists.

THE TECHNICAL SERVICE LABORATORY

The technical service laboratory exercises a most important function with respect to the overall welfare of the organization. Chemists and technicians work closely with marketing people to ensure that problems encountered with products in the field are resolved to the satisfaction of the customer. Because the work often involves problem-solving approaches, many of the procedures fall into the nonstandard category and require inventiveness and ingenuity in their application. Thus, the laboratory will generally be staffed with a higher percentage of professional and experienced chemists than one engaged primarily in control work. Also, the laboratory will usually be equipped with a wider range of analytical instruments to enable the chemist to choose an approach best suited to the solution of the problem at hand.

Problems referred to the laboratory may range from very short-range ones, where a decision is needed within a day or two, to long-range studies that may justify months of effort. In dealing with such problems chemists in this type of laboratory have a greater need to interact with peers outside the organization than do those involved in internal quality control. Well-developed communication skills may thus be particularly important for the technical service chemist.

THE RESEARCH AND DEVELOPMENT LABORATORY

This laboratory's main activity is usually devoted to working closely with other research and development personnel and furnishing them with the compositional information they require to achieve the objectives of their own work. Although standard analytical procedures are frequently used, much of the data must be obtained on materials for which precedent may be lacking. For this to be done effectively, the laboratory must be staffed by chemists who are generally more highly qualified than those in other analytical laboratories. It must also be equipped with the type of sophisticated instrumentation capable of developing the information that is needed.

Because of the availability of high-level professionals and instruments not generally found elsewhere in the organization, the research analytical laboratory may be called upon from time to time to help solve complex problems that arise in the field. Also, by virtue of its association with the development of products and processes, the laboratory may be responsible for training the staff of laboratories more directly connected with operating units of the company. Such training may consist simply of providing applicable or newly developed analytical procedures, or it may involve a period of hands-on training at either the research facility or on-site in a field laboratory.

Ideally, for the laboratory to remain technically competitive and to retain highly qualified personnel, the manager should provide ample opportunity for exploring the application of new analytical techniques for which no immediate need exists. Unfortunately, in many cases long-range benefits are sacrificed in the interests of solving existing problems.

THE ENVIRONMENTAL–INDUSTRIAL HYGIENE LABORATORY

In the past two decades there has been a tremendous increase in concern with the environment and occupational health and safety. At the same time there has been a corresponding growth in what has become known as "consumerism". This has led some industrial organizations to establish a new type of laboratory that is concerned with problems relating more directly to the health and welfare of employees, customers, and the public at large. This kind of laboratory may need to deal with a wide variety of health-related problems, including those pertaining to the assessment and monitoring of occupational hazards, toxic emissions, waste disposal, and product toxicity.

In this type of laboratory the chemists and technicians often work closely with professionals in other fields, such as environmental engineering, industrial hygiene, toxicology, and medicine. They may also be required to deal with personnel who represent various federal, state, and local regulatory agencies. Because of the nature of its activities, the laboratory will usually operate under certain constraints that may not exist in the traditional industrial laboratory. These may relate to qualifications of the staff, methods used, quality control procedures, documentation, type of reports issued, etc. As with the more conventional analytical activities, the work in this area may be carried out in a laboratory that combines several different functions.

QUALIFICATION OF LABORATORIES

In the past, with the exception of those concerned with the testing of foods and drugs, analytical chemists had few responsibilities with legal ramifications other than in the areas of patent information, contracts, or purchase specifications. Today, it is rare to find a laboratory that is not involved in some way with the Environmental Protection Agency (EPA), the Occupational Safety and Health Administration (OSHA), or other federal, state, and local regulatory agencies devoted to the public welfare. Indeed, some traditional testing laboratories may now devote a major part of their total effort obtaining information needed to conform to regulatory requirements.

In an effort to ensure the reliability of submitted data, a number of regulatory agencies have issued guidelines known as Good Laboratory Practices (GLPs).[1,2] The GLPs set forth basic criteria for qualifying the laboratory to submit its data to the particular agency involved; they relate to such items as:

- Organization and personnel
- Facilities and equipment
- Sample handling procedures
- Records and reports
- Quality control and assurance

These agencies have also established accreditation programs that require the laboratory not only to adhere to the guidelines, but to meet certain standards of performance as determined by qualification tests and periodic monitoring. Should a laboratory fail to meet and maintain the standards it may be disqualified from having its data accepted by the governmental body whose regulations it is required to meet.

In addition to the accredition systems of regulatory agencies certain others have been established in the U.S. to help validate a laboratory's qualifications. These include the National Voluntary Laboratory Accreditation Program (NIST, Gaithersburg, MD), the programs of the American Association for Laboratory Accreditation (Gaithersburg), and those of the American Council of Independent Laboratories (Washington, D.C.). National accreditation systems have now been established in scores of other countries as well. Accreditation is thus rapidly becoming a factor in world trade. International acceptance of test results obtained under comparable accrediting systems is being promoted by the General Agreement on Tariffs and Trade (GATT) and the International Laboratory Accreditation Conference.[3]

ROLE OF THE ANALYTICAL CHEMIST IN LITIGATION

Despite the fact that a laboratory may have met all qualification and accreditation requirements, its reported data are still subject to verification and challenge. In the case of contractual disagreement, problems are usually resolved satisfactorily by:

- Re-testing in the presence of the complainant's representative
- A joint analysis in one or the other of the laboratories
- Submittal of the material to an agreed-upon referee

Thus, except perhaps in support of compositional information relating to patent matters, analytical chemists have not normally been involved in legal proceedings. With the increased emphasis on regulatory testing, however, analytical personnel may now more frequently find it necessary to make depositions or even to appear in court as witnesses to support the validity of submitted data. They should be made aware of the importance of their testimony and the need for intellectual honesty consistent with their responsibility to their employer.

A discussion of the ramifications of legal proceedings is beyond the scope of this book and has been dealt with comprehensively elsewhere.[4,5] A basic consideration in the presentation of evidence is that it be technically reliable and legally sound. This is particularly important if criminal charges are involved, as in the case of certain alleged violations of EPA or OSHA regulations. If called upon to offer testimony, analytical personnel will be coached by the legal staff on matters concerning depositions and courtroom procedure. Nevertheless, they should know what criteria evidence must meet to be persuasive or even admissible.

- Witnesses must establish that they are qualified to speak on the issue by virtue of education, experience, position, and direct knowledge of the facts.
- Unless the analytical method has been specified under the terms of agreement or contract, or by a regulatory agency, it should have been one that is generally accepted. Consensus methods developed as a result of cooperative studies by the American Society for Standards and Materials (ASTM), American Water Works Association (AWWA), Association of Official Analytical Chemists (AOAC), or similar organizations are usually satisfactory; such methods are normally subjected to comprehensive peer review and evaluation for accuracy and precision before adoption.
- Testimony must be specific and pertain directly to the issue. One must avoid the temptation to make impressive-sounding comments that are likely to be ruled irrelevant. In fact, no information of any sort should be volunteered; testimony should be restricted to answering direct questions.
- The testimony should be probative and not confusing or prejudicial. For example, if the accuracy of a test result is questioned, comments relating to previous experience in similar instances are apt to be ruled prejudicial and the testimony disallowed. Likewise, hearsay and expressions of opinion will generally not be permitted.

Assuming that the above criteria are met, still other factors need to be considered. The testator should be prepared to answer certain specific questions. If the identity of a sample is involved, was the sample obtained legally, i.e., by permission or court order? How and by whom was the sample handled and was the handling thoroughly documented? Was the method followed exactly? (This question is likely to create a problem, since many methods are written to allow some leeway with respect to details that may differ from laboratory to laboratory. The chemist must be prepared to deal with any question that may arise because of differences in nonessential detail.) What quality control procedure was used, and was the method under control at the time of testing? (It is important to have documentation on hand to describe the quality control procedures that confirm the reliability of the test data.)

It is particularly important that a question be clearly understood. If not, the questioner should be asked to repeat it or rephrase it if necessary. The witness should also avoid being hurried, taking adequate time to formulate an answer, and be wary of questions that tend to shake credibility or equanimity. If a question is raised that cannot be answered, the witness should say so. This will avoid the possibility that an answer will be shown to be incorrect and thus negate the value of other testimony. One should also be alert to the possibility of being trapped by a "yes" or "no" question, e.g., "Is it possible that you have made an error?" Obviously, an unequivocal "yes" can be devastating, whereas a "no" is likely to be intellectually dishonest. Since preponderance of evidence is usually sufficient for the rendering of a decision, a question of this nature may possibly best be answered by quoting a statistical probability based on quality control data. In any case, the witness retains the right to qualify the answer, even if the examiner attempts to stifle further comment.

REFERENCES

1. CFR Title 21, Food and Drugs, Chap. 1, FDA, Part 38, "GLP's for Non-Clinical Laboratory Studies."
2. CFR Title 40, Part 792, EPA, "Good Laboratory Practices Standards."

3. **Schoch, H.E., Ed.**, *Accreditation Practices for Inspections, Tests and Laboratories*, American Society for Testing and Materials, Philadelphia (1989).
4. **Bradley, M.J.**, *The Scientist and Engineer in Court*, American Geophysical Union, Washington, D.C. (1983).
5. **Matson, J.V.**, *Effective Expert Witnessing—A Handbook for Technical Professionals*, Lewis Publishers, Chelsea, MI (1990).

2

Organization of the Laboratory

GROUPING LABORATORY PERSONNEL

A number of factors are to be considered in determining how the laboratory is to be structured. Some of these are essentially uncontrollable by the laboratory manager. For example, the organization may be influenced by the nature of the corporate business or the structure of the corporate entity of which the laboratory is a part. The economic climate, and the promulgation of government regulations that may have an impact on the nature of the laboratory's activities and its operation, also need to be considered.

One factor over which managers may have some control, although sometimes to a lesser degree than they would like, is the personnel who comprise the laboratory staff. The optimum structure needs to take into consideration not only individual technical qualifications as related to the group assignment, but personality, career interests, and personnel interactions. Age distribution also needs to be considered. It is desirable that the operating units be organized in such a way as to reflect a balance between younger staff members, who may be more knowledgeable with respect to newer theoretical concepts, and those with more practical experience. This may also minimize the dislocation that often results when a number of older staff members have to be replaced at about the same time.

It is extremely unlikely that any two laboratories will be identical with respect to the aforementioned factors or, for that matter, that a laboratory will remain the same over a period of time. Both internal and external situations change frequently, and the structure will have to be adapted to those changes. Therefore, whatever method of organizing is chosen, the probability that it will need to be changed at a future date should be taken into account.

Although it follows that the functional arrangement of one laboratory cannot be applied exactly to another, most industrial laboratories are structured according to some variation or combination of three common approaches.

By Type of Work

Where the laboratory exercises several different functions a common method of organizing its operations is to classify and assign personnel to groups according to the kind of work they perform. The grouping may be along the lines of:

- Routine testing and analysis ("standard" is a preferable term). This activity relates to test procedures that are used on a frequent and recurring basis.
- Nonstandard analyses. These are applied to samples for which precedent is lacking and which may require the use of modified standard procedures or the implementation of newly developed methods.
- Special problems. These are usually medium- to long-range studies intended to solve some problem with a product or process by the use of a variety of analytical approaches.
- Method development and analytical research. Personnel in this group investigate new procedures or improve existing ones to provide, for example, greater sensitivity, speed, accuracy, or precision. They may also conduct longer-range studies of new analytical techniques for possible application to the laboratory's needs.

This type of organization has a number of advantages, offset by possibily an equal number of disadvantages. It permits the manager to recognize individual differences in background, ability and temperament, and to place personnel in assignments that enable them to function most effectively. It also permits an orderly development of staff, allowing for progression to more advanced assignments as experience is gained. Also, as the workload changes, individuals may be shifted from one area of work to another according to their qualifications. Another advantage is that by designating a group to handle the standard testing services, particularly if similar requests are received from a number of different clients, it is frequently possible to achieve greater productivity through the economy of scale.

One disadvantage is that it is often difficult to reconcile conflicting priorities. This is particularly so in the case of standard testing for multiple clients, although conflicts can occur with the more challenging assignments as well. Another disadvantage is that it is difficult to allocate costs, particularly where method modification or development undertaken in connection with a specific client's need may ultimately be used to benefit other clients. A third type of problem may arise because of the inherent differences in the way the different assignments are handled. Those staff members who are engaged in longer-range work may spend a considerable portion of their time in a seemingly more relaxed atmosphere—in planning, library research, report writing, etc.—in contrast to those doing more active laboratory work. To avoid morale problems the manager will need to ensure that all personnel are given an opportunity to work at the level of their ability, rotating assignments if necessary.

By Profit Center

In this type of organization the laboratory is divided into units, each assigned to handle the analytical requirements of one or more process or product groups. Personnel may be shifted from one unit to another as needs vary, but basically they remain part of the dedicated unit.

One advantage of this arrangement is that it is much easier to set priorities, since conflicting pressures from multiple clients are minimized. Also, costs attributable to the process (product) are recognized more easily. A major advantage is that the analysts may, in effect, be considered part of a project team. As such, they are more thoroughly familiar with the operation and find it easier to recognize problems and contribute to their solution. It is particularly important that laboratory people in this kind of structure have the ability to interact with personnel in other disciplines.

The major disadvantage, aside from decreased flexibility in assigning personnel to cope with daily workload changes, is that it may lead to duplication of tests and equipment. Thus, while benefits to the parent operation may be greater, overall analytical costs are higher. Also, communication between members of the laboratory staff is hampered, and unless personnel are interchanged occasionally, they may become too narrowly specialized. Nevertheless, this arrangement has great merit where the increased cost can be justified on the basis of the laboratory's contribution to improved performance by the profit center. Such a satellite type of organization might be particularly attractive where the profit centers are separated physically or administratively, but where it is cost effective to have one manager responsible for the overall analytical services.

By Analytical Technique

In relatively large laboratories, particularly those concerned with multiple functions, a common method of organizing the staff is to assign personnel to groups, each dealing with a particular type of analytical methodology. Such groups might be designated, for example, as wet-chemistry elemental analysis, organic functional group analysis, X-ray spectrometry, gas chromatography, mass spectrometry, etc. Depending on the size and nature of the operation, a senior chemist may be responsible for one or more of these groups under broader designations, e.g., wet chemistry, instrumental analysis, physical testing.

As with the other organizational structures there are advantages and disadvantages. A major advantage is that it avoids costly duplication of facilities. Another is that the group develops a specialized knowledge of the problem-solving capabilities of the various techniques involved. There is also greater flexibility within the group. As the workload changes junior staff members can be assigned to either standard testing or the support of senior personnel engaged in nonstandard analytical activities.

One disadvantage is that with routine service work and longer-range activities being conducted in the same group, the question of priorities with respect to both staff and equipment utilization almost always arises. Managers and group leaders will need to exercise sound judgment in utilizing available resources and, at the same time, maintaining group morale. Another disadvantage is that the chemists may become too narrowly specialized. This will not only limit the manager's flexibility in making temporary reassignments to handle day-to-day workload changes, but it will also deny the specialist an opportunity to develop first-hand familiarity with other areas of the laboratory's activities. Thus, chemists in the specialist group may attempt to solve problems by the technique they are most familiar with, even though another approach might be preferable. The manager can help overcome this disadvantage by avoiding too narrowly specialized groupings and making periodic rotational reassignments.

To be able to make reasoned decisions, potential managers should have sufficient knowledge of the capabilities and limitations of various techniques; therefore, a senior staff member who has served in a specialist assignment for an unduly long period may not be as suitable a candidate for promotion to managerial status as would a generalist. In any organization managers can help prepare for their own advancement by developing qualified replacements. If a specialist has indicated interest in advancing along managerial lines, the manager should make a particularly strong effort to broaden his or her background by periodic rotational reassignments.

The Open Laboratory

In recent years, a number of academic institutions have installed what is known as an "open" analytical laboratory.[1] In this type of laboratory other members of the institution are authorized to use certain instruments, on an availability basis, for making their own analytical measurements. The analytical staff is primarily responsible for maintaining the equipment and instructing others in its use. In consideration of the high cost of instruments, this arrangement has merit in avoiding duplication of facilities.

The practice does not seem to be widespread in industry, although there are instances where research scientists prefer or even insist on making their own measurements. Delays in acquiring needed information may be given as a reason, but personal factors are probably more often a greater consideration. A basic requirement is that the user should be as qualified as the analytical chemist to evaluate the reliability of the data obtained. Although there may be exceptions, this is not likely to be the case with the occasional user as opposed to one who may use the instrument on a regular basis and is familiar with the vagaries of the procedure and the instrument. In the author's view, the practicality of this approach on a regular basis, other than in an academic environment where the focus is on training, is questionable. If the concept is adopted, use of the laboratory's instruments by nonanalytical personnel should be restricted to those occasions when there may be compelling advantages to the organization as a whole. Also, a charge-back system for instrument time should be adopted to defray part of the overhead costs and to discourage capricious use.

As already observed, selection of the organizational structure depends on the individual situation. Large laboratories, in particular, will often use combinations of the above arrangements. Cook[2] has described a number of functional configurations and suggests that no matter how they are structured for service analyses, some laboratories may find it advantageous to have part of their staff dedicated to performing regulatory test work. The value of this activity, however, cannot be judged on the same "profit or loss" basis as others and must be looked upon as an unavoidable cost of remaining in business.

Regardless of how the staff is organized there are several basic considerations. One is that the organization be under constant review. As staff changes occur, it is important to recognize individual abilities and personal characteristics by possibly reorganizing, i.e., fitting the organization to people rather than people to the organization. Another consideration is that an operating unit should have a sufficient minimum number of qualified people assigned to it so as to reduce the disruptive effect of absences; this will avoid the need for frequent temporary reassignments. Finally, at least in the category of direct service, it is desirable to solicit the views of the clients to determine if their needs are being best met by the organizational structure of the laboratory. Although different clients may offer conflicting opinions, the fact that the laboratory is solicitous of their views cannot but help peer relations.

LAYOUT OF THE LABORATORY

Managers rarely have the opportunity to design the physical facility for which they will be responsible. In many instances they will have inherited a plant that from a physical standpoint is probably far from ideal, and the laboratory arrangement must be determined by the structure of the building and the amount of space available. Within these constraints the laboratory may be arranged in several ways.

Open Arrangement

In this arrangement essentially all of the analytical work is carried out in one large laboratory room. Separate areas may be designated for the different types of testing, e.g., physical, wet-chemical, and/or simple instrumental techniques. This arrangement is usually found only in older and smaller control laboratories in which the work is primarily routine in nature and the staff consists largely of low-level professionals and technicians. For this purpose, it may offer the advantage of facilitating communication and supervision. It also permits community use of certain equipment and supplies such as ovens, balances, reagents, etc., thereby minimizing duplication. There are, however, several offsetting disadvantages. Some instruments may suffer from exposure to an unfavorable environment. Housekeeping and maintenance may suffer as a result of shared responsibility. Also, personnel are likely to be more easily distracted by conversation and discussions unrelated to the technical activities.

Compartmental Arrangement

In this type of laboratory layout the work is carried out in a number of smaller rooms, each dedicated to a different activity.

Although it has certain disadvantages of its own, it overcomes some of the defects of the open arrangement. It is far more appropriate, even essential, where the staff includes higher-level professionals and where the technical activities involve the use of modern instrumentation. Morale is improved, and there is less of a problem in providing a suitable environment for the instruments or methods that are used. However, the advantages of the open arrangement with respect to supervision, communication, and avoidance of duplicate facilities become disadvantages in this case.

Mixed Arrangement

In the majority of cases, and certainly if the laboratory is responsible for a wide range of testing and analytical functions, a combination of the above two types of layout will be found to be most effective. Activities such as those requiring close supervision of a number of technicians performing related assignments are conducted in an open section of the laboratory. The remaining space is allotted to those activities for which a compartmental arrangement is most suitable and is partitioned according to the particular functions to be performed.

A very simple form of this type of arrangement is illustrated in Figure 1. Large laboratory sections are designated for several types of standard testing activities, with an adjacent office area for the group supervisor. The required number of smaller laboratory and office areas are reserved for other activities, such as those utilizing instruments that are best isolated, laboratory and office facilities for higher-level professionals, sample receiving, sample preparation, "clean" or climatically controlled operations, sample retention, library, conference room, etc. Where smaller office or special purpose areas (e.g., balance room, sample preparation) are desired they may be installed between laboratory areas and shared if necessary.

As in the case of organizing personnel, one must consider the probability that the spatial arrangement will have to be changed at some point. To the extent possible, a modular structure should be chosen to minimize the problem of relocating partitions and utilities.

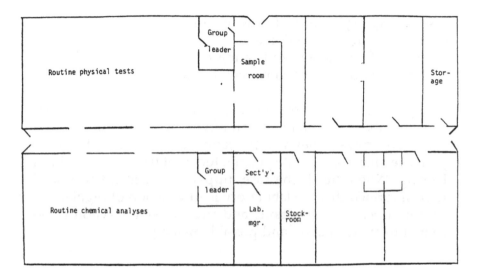

Figure 1. Simple laboratory layout.

PHYSICAL FACILITIES

As noted, laboratory managers have not always been able to influence the design of the facility which they will have to operate. It may also lack certain features that the manager considers essential and, especially if it is an older installation, it may fail to meet certain OSHA[3] or EPA requirements.[4] Although the mandatory or desirable features will depend on the particular activities of the laboratory, most are basic to almost every laboratory. These include:

- An adequate number of hoods, all equipped with a performance monitoring device
- Stockroom for chemicals and supplies
- Spare equipment storage facility
- Adequate space between work areas
- Convenient and safe storage and disposal facilities for volatile and flammable materials, and toxic waste
- Eyewash fountains and drench showers (with drain)
- Multiple exits from laboratory areas
- Alternate power source for critical operations
- Fail-safe escape mechanism from cold room
- Facilities for expansion of computer lines and networking

- Vented gas cylinder and solvent storage areas
- Fire extinguishing equipment
- Separate food storage and eating facility
- Cloakroom or lockers conveniently located

Experienced laboratory managers will undoubtedly recognize the above requirements and even be able to add to the list. Nevertheless, they will probably confirm the author's experience that some of the mandated facilities are sometimes lacking. Where this is the case, upper management will need to be informed that the laboratory is in violation of regulations on safe operating conditions and that corrective action is essential to avoid citation and possible penalty.

REFERENCES

1. **Lucchesi, C.A.,** The analytical services laboratory at Northwestern, *Am. Lab.* **13**(7):24 (1981).
2. **Cook, C.F.,** Analytical organization and design in a changing environment, *Anal. Chem.* **48**(9):724A–738A (1976).
3. CFR Title 29, OSHA, Part 1910, "Occupational Exposures to Hazardous Chemicals in Laboratories."
4. CFR Title 40, EPA, Part 262, "Standards Applicable to Generators of Hazardous Waste."

3

Staffing the Laboratory

One of the manager's most difficult challenges is the recruiting and selection of a competent staff. Although there have been occasional periods of economic downturn when demand has slackened, the general trend has been one of increasing opportunities in the field of analytical chemistry, particularly for high-level professionals.

Until about the mid-1950s relatively few chemists performing analytical work held doctorate degrees. In fact, records of the National Research Council show that during the immediate post-World War II period the number of doctorates awarded in analytical chemistry had dropped to about 1% of the total doctorates awarded in all the other branches of chemistry. Indeed, a number of prestigious universities did not offer the degree in analytical chemistry, and at the undergraduate level some schools dropped separate courses in the discipline and incorporated the subject matter into other course offerings.

By the late 1950s a perceptible change, attributable to two major factors, occurred. First, there was a growing recognition of the importance of minor and trace concentrations, speciation, and molecular structure. The development and use of the sophisticated techniques needed to obtain this type of information required a stronger academic background than had formerly been considered necessary. Second, and possibly more important, was growing concern with the environment, followed by the establishment of the EPA and issuance of the first significant federal regulations dealing with air and water pollution. This focused on the need for a comprehensive evaluation of existing analytical technology and, where it was found

Table 1. Chemistry Doctorates Awarded by U.S. Universities*

Year	Ph.D. Degree in Analytical Chemistry	% of All Ph.D.s in Chemistry
1975	142	8.0
1976	152	9.4
1977	174	11.1
1978	178	11.5
1979	207	13.2
1980	184	12.0
1981	229	14.2
1982	190	11.3
1983	264	15.0
1984	228	12.9
1985	285	15.5
1986	257	13.8
1987	314	16.2
1988	301	14.9
1989	289	14.7
1990	293	13.9

*Includes foreign nationals.

Source: National Research Council, Doctorate Records File.

lacking, aggressive method development by highly qualified scientists had to be undertaken. Thus, government and industry joined with academia in competing for analytical chemists at the doctorate level, and a career in the field became more attractive.

AVAILABILITY OF CAREER ANALYTICAL CHEMISTS

Table 1 shows the annual number of analytical chemistry doctorates earned in U.S. univerities since 1975 and as a percentage of all chemistry doctorates. Despite the gradual improvement during the past decade and a half, indications are that the number of Ph.D.s has peaked; moreover, the number of foreign students among them is increasing. During the decade of the 1980s, the percentage of foreign students in graduate chemistry programs continually increased to reach a total of 33% by 1989.[1,2] Under present immigration law these stu-

Table 2. Degrees Awarded in Chemistry—
1980 to 1990*

Year	Bachelor's	Master's	Doctorate**
1980–81	11,347	1,654	1,613
1981–82	11,025	1,618	1,683
1982–83	10,796	1,622	1,759
1983–84	10,704	1,667	1,777
1984–85	10,482	1,719	1,806
1985–86	10,116	1,754	1,885
1986–87	9,661	1,738	1,936
1987–88	9,025	1,694	2,018
1988–89	8,654	1,785	1,971
1990***	7,650	1,605	2,102

*Source: U.S. Department of Education, Digest of Education Statistics—1990.
**Source: National Research Council, Doctorate Records File. Reported on calendar year basis.
***Source: ACS Committee on Professional Training (ACS approved programs only; Bachelors approximately 92% of all Chemistry Bachelors).

dents, with rare exception, may not be employed in the U.S. once their studies are completed.

Another measure of the problem facing the manager seeking to employ highly qualified professionals may be obtained from the "Academic Positions Open" listings in *Chemical and Engineering News*. The author has surveyed the issues from July 2 to December 24, 1990 for the number of tenure-track vacancies where a specific chemistry discipline was designated. A total of 496 (not edited for repetition in different issues) was listed under six broad areas of specialization. Of these, 106 were for analytical faculty. Industrial openings were not surveyed since those are frequently described in more general terms, but it seems reasonable to assume that the relative number of openings for analytical doctorates in industrial and other laboratories is similar to that in academia. Thus, managers seeking to recruit doctorate-level analytical chemists are likely to face continuing difficulties.

The problem exists at the Bachelor's level as well. Table 2 shows the total number of chemistry degrees awarded at the three academic levels during the decade of the 1980s. It is clear that although an increasing number of higher degrees were earned, the number of students completing a basic chemistry curriculum has been decreasing steadily. The 1990 total rep-

Table 3. Undergraduate Students in U.S. Colleges

Academic Year	First-Year Enrollment*	Fourth-Year Completions
1980–81	1,646	935,140
1981–82	1,568	952,988
1982–83	1,562	969,510
1983–84	1,662	974,309
1984–85	1,539	979,477
1985–86	1,499	987,823
1986–87	1,503	991,339
1987–88	1,575	993,362

*In thousands.

Source: U.S. Department of Education, Digest of Education Statistics—1990.

resents a decrease of over 25% from that of 10 years earlier. Thus, the smaller pool of available undergraduates upon which to draw for entry level appointment or for graduate study is not likely to match the needs of an expanding technological economy.

Table 3 shows the number of all college entrants and fourth-year completions during much of the same period. A comparison of the data shows that there has been an increasing number and percentage of completions. Thus, not only has the absolute number of chemistry graduates decreased, but so has the percentage relative to all other college graduates; the data show that it is now under 1%. The problem of too few students choosing to study chemistry exists in Canada as well.[3] It is also noteworthy that the American Chemical Society has found that of the 1990 chemistry graduates, almost one third plan to pursue graduate study in fields other than chemistry, mostly medicine and related professions.[4]

The scarcity of chemistry-oriented students is particularly significant from the standpoint of the analytical laboratory manager; an earlier ACS survey[5] found that 20% of all chemists listed analytical as their work speciality—substantially higher than any other group. It is obvious that unless the present trend is reversed, analytical laboratory managers will in the future face even greater problems in staffing their laboratories with chemists trained in the discipline.

Further evidence of the problem facing the laboratory manager is found in data from the Employment Bureau of the Pittsburgh Conference. From 1984 through 1990 a total of 5899 employers and 5073 candidates were registered;[6] in some years employers actually outnumbered candidates. If one takes into consideration the fact that some employers listed multiple openings, while of those registered as seeking employment many were no doubt already employed and seeking to improve their status, the figures become quite significant.

PROMOTING ANALYTICAL CHEMISTRY AS A CAREER

A number of approaches have been suggested to help ensure a greater availability of qualified staff. On a general level, the Analytical Division of ACS has for some years operated several student programs. In addition to providing graduate fellowships, it offers summer internships, primarily for graduate students majoring in analytical; other eligible students are those who have completed a minimum of 2 years of college and are considering analytical as a career. Interns have been placed in industrial, government, and academic laboratories. To stimulate interest in the field, the Division may also award meritorious undergraduate students honorary membership and a subscription to the journal, *Analytical Chemistry*.

Laboratory managers may take independent steps to encourage interest in analytical chemistry among college chemistry majors and perhaps at the same time recruit future staff members. One way is to institute a program in which laboratory personnel address students, pointing out the interesting and rewarding aspects of analytical laboratory work. One such program is the short course presented by The Procter and Gamble Company to upper-level undergraduates to encourage them to seek a Ph.D. degree in the field.[7] The course focuses on the role of the analytical chemist as a problem solver, method developer, and consultant—concepts that are very different from those that the student may have developed in undergraduate studies. By the use of case histories describing the role of the laboratory in solving problems, it has proved to be quite successful in stimulating interest in analytical chemistry as a career. Similar presentations, geared to the level addressed and utilizing examples from one's own laboratory experience, could prove equally effective.

Another step that has proved useful in promoting interest in analytical chemistry is the use of students for temporary employment during summer vacation periods. The nature of such programs varies. Some are intended for college seniors who plan to enter graduate school; these students may be assigned to work on a research project, either independently or with a member of the permanent analytical research staff. If this type of program is instituted it is desirable that, insofar as possible, the project be one that can be expected to be completed within the period of the student's tenure to promote a sense of accomplishment. In any case, the temptation to use the student for routine testing must be avoided, lest the effort prove counterproductive by distorting the true nature of professional analytical chemistry.

Other summer programs are intended for students who are less advanced but who, it is hoped, can be encouraged to develop an interest in the field of analytical chemistry. Although assignments with senior chemists handling research-type problems may also be found for such students, they are usually given more routine tasks. Regardless, the student should be given ample opportunity to become familiar with the type of problem that the professional analytical chemist deals with and learn the importance of the laboratory's contribution to the overall welfare of the organization.

Still another way in which the laboratory can encourage students to seek careers in analytical chemistry is to participate in a cooperative education program. In recent years there has been a significant increase in the number of institutions offering cooperative programs. The American Chemical Society's Office of Cooperative Education has identified several hundred schools in the U.S. and Canada that offer programs in chemistry;[8] included in the list are Associate as well as Graduate programs. The periods allotted to employment in industry vary with the institution, but in a cooperative program the laboratory is not restricted to using students only during the summer vacation period.

Table 4. Salaries of Analytical and
Organic Ph.D. Chemists*

Years Since Bachelor Degree					
5 to 9	10 to 14	15 to 19	20 to 24	25 to 29	30 to 34
		Analytical Chemists			
46.0	51.2	58.0	62.0	65.4	67.1
		Organic Chemists			
45.7	51.0	57.0	65.0	70.0	72.5

*Median salary, in thousands (as of March 1, 1990).

Source: ACS salary survey, *Chem. Eng. News,* July 9, 1990.

Aside from emphasizing the professional satisfaction of a productive career in analytical chemistry, laboratory managers can do much to enhance the status of their staff by continuing to ensure that upper management is aware of the importance of maintaining an equitable salary structure. In past years, analytical chemists often lagged beyond their peers in this respect. This situation seems to have improved. Table 4 lists median salaries for analytical and organic Ph.D. chemists. In general, it may be concluded that analytical chemists who obtained their degrees within the past 20 years have not suffered by comparison. Similar data at the Bachelor's level also show that the salaries of relatively recent graduates who are in the field of analytical are comparable to those of organic chemists.

RECRUITING PERSONNEL

Technicians

Nonprofessional technicians may be recruited in several different ways. In many instances, the laboratory is part of a larger industrial organization in which qualified personnel may be employed in less desirable positions. Posting of an opening on bulletin boards or otherwise publicizing it will often result in applications from suitable candidates. Even if not, demonstration of the laboratory's desire to promote from within will have a beneficial effect on employee morale.

Often, the laboratory will be subject to a collective bargaining arrangement. If so, the agreement will usually *require* that vacancies in the nonexempt category be posted for bids by eligible employees. In some cases, bids under such an arrangement are received from applicants whose education or background is such that their ability to meet the minimum job requirements is questionable. It has been the experience of some laboratories operating under such constraints that the manager must accord more weight to seniority than to qualifications, unless the latter are *clearly* inadequate. This may be difficult to prove without a costly trial period. In some cases, selection may be made from among several of the most senior applicants, with a provision that a candidate who meets the minimum requirements must not be bypassed more than some prescribed number of times without a trial. If a manager adds a staff member in this manner and subsequently claims that performance has been unsatisfactory, it will usually be necessary to show that every effort was made to help the employee do a satisfactory job.

A second source of technicians is the 2-year community college or technical school, which may offer an Associate degree. Many managers have reported on the excellent background provided by such institutions and the minimum amount of on-the-job training needed before such staff additions become proficient in carrying out their assignments. Laboratories located near such a facility are especially apt to find it a productive source, since persons at the technician level are less inclined than professionals to make a major geographical change. Also, such institutions welcome the participation of people from industry in an advisory capacity. If the manager or a senior aide serves in this way, the laboratory cannot only influence the course content but probably also gain an advantage in recruiting the most promising students.

Personnel placement agencies and newspaper advertisements are sometimes used to recruit technician (less often for professional) staff additions. Experience has been mixed, however. Although some managers report that they have employed qualified candidates recommended by employment agencies, many have found that certain agencies tend to submit resumes that have little bearing on the qualifications sought.

Where agencies are used, requirements should be clearly stated, and it should be emphasized that any indiscriminate submittal of resumés will result in disqualification of that agency from further consideration. Similarly, if advertisements are placed in the newspaper, qualifications and job location should be stated so clearly that the need to review (and respond to) a large number of applications from obviously unsuitable candidates will be minimized.

In some cases professionals at the Bachelor's level are used as technicians. This practice is not generally considered acceptable, unless the assignment is for training or development purposes. If so, the intent should be made clear to the employee and the assignment should not be prolonged unduly. It is doubtful that professionals of the caliber desired in modern-day analytical laboratories will be content to perform repetitive tasks for an indefinite period.

Professionals

Beginning professional chemists are usually recruited as a result of on-campus visits and interviews. Most large organizations have a college recruiting staff which schedules and conducts preliminary screening interviews for different segments of the company. Candidates judged favorably on the basis of the interview and/or faculty recommendations are then referred to the appropriate department for further consideration.

Although college recruiting may be utilized for all academic levels, laboratories seeking to recruit professionals, particularly at the doctorate level, often assign a senior staff member to conduct on-campus interviews. These may be handled independently or in cooperation with the corporate recruiter. The participation of a chemist in the interview process permits a better preliminary evaluation of the candidate's technical background; it also enables the interviewer to answer specific questions relating to the nature of the work being performed at the laboratory. If available on the staff, a qualified graduate of the institution is a preferred choice for the assignment.

Professionals are also frequently recruited from registers maintained by national technical societies or their local branches. Such organizations as the American Chemical Society, the Pittsburgh Conference, and the Eastern Analytical Symposium also conduct employment bureaus in conjunction with their meetings. The registers may be used both to list openings and to consult for resumés of those seeking employment. At meetings, interviews may be scheduled on-site or the resumés may be referred to for subsequent action.

Both the "Situations Wanted" and the "Positions Open" sections of *Chemical and Engineering News* provide an opportunity to locate potential additions to the professional staff. Users should be aware that it is considered obligatory for all replies to be acknowledged; therefore, advertisements of openings should be as specific as possible to reduce the number of applications from unqualified persons and the need to respond to them. Also, the company and the location should be clearly identified; box numbers are unsatisfactory. Most desirable applicants are not likely to respond to "blind" ads, while others may have geographical limitations.

Temporary Employees

Many laboratories find the use of temporary employees to be very helpful during periods of peak loads or when absences are high, for example, during the summer vacation period. Students have been commonly used for this purpose. In some cases students have also been used on a more or less regular part-time basis. As with co-op students this provides the added benefit of allowing the manager to evaluate the student for possible eventual appointment to a permanent position.

In recent years, some laboratories have employed for temporary or part-time work in a technician type of assignment former employees who may have elected early retirement. This practice has the obvious advantage of having an experienced analyst who can become immediately productive without the need to spend time for additional or specialized training. Managers have commented on the enthusiasm often displayed by such employees, who enjoy the freedom of a part-time schedule that still provides the satisfaction of gainful employment.

DESIRABLE PERSONAL CHARACTERISTICS

An adequate educational background is rarely sufficient to

ensure that a candidate will become a productive member of the group; personal qualities are in many cases equally or more important. The particular qualities that one seeks in candidates for laboratory work are in many respects quite similar, whether it be in analytical or any other branch of chemistry. They are also similar regardless of whether the applicant is a professional or technician, although different weight may be assigned to certain characteristics, depending on the level and nature of the vacancy to be filled.

Most managers would probably agree that among the qualities one would want technicians to possess the following are most important:

- Cooperation
- Ability to learn readily and follow directions
- Observant of details
- Ability to communicate orally
- An interest in physical science
- Good number sense
- Ability to work with hands

With the possible exception of the last of the above qualities, these are equally important for the professional employee but, for the most part, may be taken for granted. There are, however, others that may be considered even more essential. These are

- Inquisitiveness
- Critical of detail
- Resourcefulness
- Adaptability
- Originality
- Ability to convert theory to practice
- Ability to communicate effectively in writing as well as orally
- Ability to work cooperatively with subordinates, superiors, and peers
- Ambition to advance

In addition to the above qualities, which would be desirable in any scientific or technical field of endeavor, professional analytical chemists must be inclined to be perfectionists. At the same time, they must have the judgment to recognize that there are occasions when the highest degree of accuracy is not required, and that continued costly effort may be counterproductive. For example, it is often sufficient to characterize a component as being above or below a prescribed level without specifying an actual concentration. A common difficulty managers face is how to handle the employee who stubbornly wishes to continue working on a problem when it is no longer cost-effective to do so. (See "Case History—The Perfectionist", Chapter 4.)

Most important of all, professionals must look upon themselves as "analytical chemists", not "chemical analysts". Recognition of the difference is essential. They must recognize that analysis of the sample is not an end in itself, but that the sample is merely representative of a problem; the laboratory's responsibility is to determine what contribution it can make to the solution of that problem.

In describing analytical chemistry, Laitinen[9] has decried the archaic conception—"The traditional role of the analytical laboratory . . . is to receive samples, execute measurements (perhaps as specified by the customer) and report results." According to this concept, the laboratory is looked upon as a "black box", operated by chemical analysts, into which one introduces a sample and from which numbers emerge. Analytical chemists, on the other hand, are problem solvers who, as described by Thorpe,[10] " . . . utilize analytical thought processes, specialized knowledge of analytical measurement techniques, and broad training in the chemical sciences . . . to characterize experimentally chemical systems in qualitative and quantitative terms."

REVIEWING JOB APPLICATIONS

Resumés will often provide sufficient information to allow the manager to decide on the advisability of an interview. However, it should be kept in mind that they are sometimes prepared under professional guidance in such a way as to create the most favorable impression and encourage issuance of an invitation. Clues to possible problems with a candidate are often provided by considering the following:

- Is it neatly prepared and well-organized?
- Is it reasonably complete?
- Is good judgment shown in the way questions are answered?
- Does it show evidence of claims inconsistent with the experience level of the applicant?
- If the applicant has been employed previously, were there frequent changes of employment? If so, were the moves lateral or were they advancements?
- Are negative comments made about previous employers? Are the reasons for leaving inconsistent with what is known about those employers' policies?
- Are there unexplained lapses in the applicant's personal history or periods shown as "self-employed" or "consultant"?
- If the applicant is a professional, are there errors in spelling or grammar that suggest carelessness or inability to communicate effectively in writing?

Where additional insight into detailed accomplishments or personal qualities is desired prior to (or even after) the interview, previous supervisors may be consulted if confidentiality has not been requested. However, one often needs to evaluate the reliability of the information obtained. A current employer, with whose knowledge the applicant is seeking new employment, will possibly give a marginal employee a favorable recommendation. Similarly, if the candidate has been or is about to be terminated, the employer may feel some obligation to help that person find employment elsewhere and disguise the reasons for the termination.

In many cases, to avoid possible legal entanglement by commenting on previous employees, employers will only confirm dates of employment and position held. For the same reason, if information is given it is more likely to be given on an informal basis rather than in writing, particularly if it includes some negative comments. It is suggested that requests for information are more apt to be honored if worded, not as a request for a recommendation, but rather in such a way as to indicate only that confirmation of specific points is desired.

If available, faculty contacts are usually consulted for recommendations on graduating students or recent graduates. Especially in large institutions, however, they may not have sufficiently intimate knowledge of their students, and judgment may be based solely on grades or performance on a narrowly specialized research project. In any case, grades alone may be particularly misleading, since they may have been affected by factors completely unrelated to academic ability. Highly motivated students, for example, may have worked long hours on outside jobs to cover the costs of their education.

Other applicants may be acquainted with personnel presently on the staff and who would become co-workers if the appointment is made. Such staff members are probably among the best sources of information, but here, too, the nature of the relationship should be considered to rule out possible bias based on personal friendship.

INTERVIEWING CANDIDATES

Preparing for the Interview

The success of the interview depends to a great extent on the degree to which the manager has planned for it. It is essential that a schedule have been prepared, especially for professional candidates, who normally also meet with key staff members and potential colleagues. A haphazard effort in trying to arrange meetings with others after initial discussion with the manager is likely to create an extremely unfavorable impression. The preparation should also have included a clear delineation of who will explore what areas; the candidate will not then be irked by repetition of questions and the information gained will be more comprehensive. It is also necessary that all interviewers be aware of what questions may properly be asked and which ones may lead to legal difficulties under various state and federal antidiscrimination statutes, particularly if the candidate is rejected. For example:

- Candidates may not be asked their age.
- One may discuss schools attended and diplomas or degrees obtained but may not ask when, lest it be interpreted as an attempt to establish age.

- Inquiries may be made regarding competence in a foreign language, but asking how it was obtained might be construed as an attempt to determine ethnic background.
- A question regarding citizenship is proper, but one to determine whether it was acquired by birth or naturalization is not.
- Questions relating to a female's marital status or her plans for raising a family are not only illegal but insulting.
- It is proper to ask whether the applicant has ever been convicted of a crime, but not whether ever arrested.

The interviewer should also be aware that illegal information of the above nature, even if not solicited directly, may prove embarrassing to the company if it can be claimed that the applicant was led into volunteering the information. Unsolicited comments relating to these subjects that may be made by the applicant should be dismissed by the interviewer as having no bearing on the position.

Conducting the Interview

It is important to realize that as far as the applicant is concerned the interviewer is the organization. Therefore, and particularly if the company is consumer-oriented, it is essential that a favorable impression be made, even if it becomes evident that an employment offer will not be made. A candidate for employment should be made to feel that the interview is equally important to the laboratory. Arrangements should be made to avoid interruption by visitors or telephone calls. Equally disturbing to an applicant is the repetition of questions already answered by the resumé; this will create the impression that not much thought was given to the information provided. Notes on the important items to be discussed should have been made in advance.

Testing is not normally done either during or prior to the interview, except possibly for technician openings. Professionals may be given a problem that the laboratory has solved and asked how they would handle it, but this should be done within the framework of a description of the laboratory's function. To provide an indication of breadth of knowledge, a problem that required the use of analytical techniques outside the candidate's area of specialization might be chosen.

Many managers prefer, at least for professional applicants, an informal "chatty" type of interview in which the candidate is encouraged to do most of the talking. A "one-on-one" setting helps the interviewee feel at ease and perhaps more forthcoming, even though several interviewers may be involved during the course of the visit. Laboratory managers have expressed mixed opinions on whether notes should be taken during the interview or not; if they are, it is recommended that they be jotted down casually to enhance the atmosphere of informality. In any event, it is essential that complete notes be prepared immediately afterwards while the impressions are fresh. Robert Half, a prominent management consultant, has found that the last person interviewed is three times as likely to be hired as earlier interviewees; he concludes that this is primarily because the interviewer has failed to recall the qualities of the others, making the last one seem superior.[11]

With professionals, the interviews by senior staff members are usually carried out in the course of visiting various parts of the laboratory with the candidate. This not only encourages informality but also provides a better opportunity to judge the candidate's technical background, inquisitiveness, and attention to detail. Candidates who have recently acquired, or are about to acquire, the Ph.D. degree are also often asked to present to a brief seminar on their doctoral thesis. The clarity of the presentation and the manner in which questions are handled provide an excellent opportunity to evaluate communication skills, interaction with peers, and perhaps response to stress.

Candidates, especially those from institutions that have well-organized placement offices or who have been through a number of interviews, are often primed to answer certain specific questions; distinguishing between a sincere answer and a rehearsed one may be difficult. Hence, lists of specific questions that are sometimes recommended in interview-training courses are not always helpful, and the candidate who has the ready "right" answer may not be as attractive as one who hesitates and is somewhat uncertain.

Questions that can be answered by "yes" or "no" or by a

brief statement should avoided as being unlikely to elicit the kind of information that is desired. A preferred approach that encourages the applicant to be forthcoming is to minimize short direct questions in favor of phraseology such as "You say that you ____" or "I see from your resume that you ____" or "please tell me as much as you can about why ____, when ____, how ____." What and how much is said in reply and how it is expressed usually permit the manager to better judge how well the candidate will meet the requirements of the position than do the brief answers sometimes unwittingly encouraged.

Conclusion of the Interview and Decision

Interviews are normally concluded by determining if the interviewee has any questions. The number of questions asked and their nature can provide considerable insight into the character of the candidate. For this reason, it is desirable to almost force questions by some phrasing along the lines of "We've been talking mostly about you. What can I tell you about our policies?" The candidate who asks questions about developmental opportunities, participation in decisions, freedom to exercise initiative, etc. is normally preferable to one whose questions deal with benefits such as vacation policy, sick leave, bonuses, etc. In the final analysis, the interviewer will usually have a "gut feeling". If there is some element of concern, even if it can't be explained, it is probably best to reject the applicant.

During the course of the interviewing period (usually a full day in the case of Ph.D. candidates), the manager has presumably had an opportunity to review the opinions of the other interviewers. An immediate formal decision is not often made, except in the case of a very strong or weak prospect. If the consensus is favorable and an offer is likely to be forthcoming, the candidate's interest in the position should be determined. At the doctorate level at least, the usual response is that an offer would be welcomed and considered. Although this answer may be noncommittal, the manner in which it is expressed and the enthusiasm shown often provide a clue as to whether it is a sincere expression or merely a polite recognition of the time and effort expended by the laboratory personnel.

The manager will thus be guided in making a decision on what formal offer(s) will provide the best prospects of meeting staff requirements.

Whether it is probable that an offer will be made or not, the prospect should be told by what date a decision can be expected. It is essential that the date be met. Occasionally, even if the decision is positive, some unforeseen development makes it impossible to communicate the formal offer by the stated date. For example, it may require the approval of a higher-level manager who is temporarily unavailable. In that case, the candidate should be informed and given the opportunity to indicate whether further delay would cause undue hardship. This may forestall the candidate's acceptance of a second-choice offer rather than risk being rejected. The same approach may be used even if events preclude a firm decision. If the decision turns out to be negative, the fact that the manager has met the commitment will avoid leaving the candidate with an unfavorable impression of the company.

ADDITIONAL CONSIDERATIONS IN STAFFING

Increasing numbers of women are entering all branches of science and engineering, including analytical chemistry. In line with this trend, there has been a great increase in the number of dual-career professional couples. On occasion, acceptance of an employment offer has been contingent on a spouse finding suitable employment in the same geographic area. Managers have thus been forced to deal with a situation in which an attractive prospect may reject an offer in favor of employment elsewhere, regardless of how mutually beneficial the association might seem otherwise.

If the laboratory is part of a large company and the spouse is qualified in an area of interest, the manager may find it possible to have employment offered to both. Another possibility is to solicit the aid of the company personnel division. Intercompany affiliations may be helpful in such instances, and the personnel director may be able to arrange for employment to be offered to the spouse by another company in the same general locale. Even if the attempt to help is unsuccessful, a realistic recognition of the prospective employee's problem may result in the offer being considered more favorably than a competitive one.

Another consideration, applicable to the laboratory along with other branches of the company, is the growing practice of granting emergency leave to employees for child or parental care in case of serious illness. Although a 1990 attempt to require this under U.S. federal law failed by a narrow margin, new legislative attempts are being made. Also, competitive pressures may force organizations that do not have such a policy to adopt a more liberal approach to the problem. To cope with the possibility that a member of the laboratory might have to be granted leave for an extended period, the manager should have a plan in place for the use of temporary replacements in case of need.

ORIENTATION OF THE NEW EMPLOYEE

General company policies and the procedure for the orientation of new employees are usually clearly described by the company personnel department. In the case of the laboratory, certain other steps may be taken to help the new employee become a productive member of the group.

First and foremost is the matter of safety. The topic is discussed more fully in Chapter 6, but it is essential that the manager emphasize early the importance that the laboratory places on safety. This should be done in the course of outlining laboratory policies when employees first report for duty. Prior to the more formal training required by OSHA[12] every new employee should be given a copy of the laboratory safety manual and thoroughly informed on the laboratory's general safety rules and practices by the safety director or, in the absence of one, the head of the laboratory safety committee. Finally, the employee's immediate supervisor should then review the specific safety factors that relate to the employee's assignment. It is recommended that a form be adopted to be signed by each of the persons involved, including the new employee, attesting to the completion of the above steps; this should then be placed in the employee's personnel file.

A good practice is to give all new staff members, regardless of status, a copy of the laboratory organization chart. A clear understanding of where co-workers fit into the scheme of things can do much to minimize the confusion that every new employee encounters during the first days on the job. Professional employees, depending on their level, should receive appropriate copies of other organization charts. These might include, for example, charts of those client sections that generate samples and problems for the laboratory to handle. Highest-level employees might also receive a chart that shows how the laboratory relates to broader segments of the company.

The numerous orientation details bearing on job-related facilities are left for the new employee's immediate supervisor to explain. However, the manager should take the opportunity to outline for professionals the laboratory's procurement procedures and define the employee's limits of authority to commit funds for materials, services, supplies, etc.

As part of their orientation, new high-level professionals should also be given an opportunity to spend some time visiting with various client groups to become familiar with their personnel and activities. In the course of such visits, views might be solicited on how the laboratory can be of greater assistance in meeting the clients' needs. Such discussions may not only result in helping the laboratory utilize its efforts more effectively; they can also be used to promote the concept that the laboratory is not simply a facility to generate data but rather a problem-solving adjunct to the company units it serves.

REFERENCES

1. *Chem. Eng. News,* Oct. 23, 1989, p. 30.
2. National Science Foundation, reported in *Chem. Eng. News,* Aug. 20, 1990, p. 52.
3. *Chem. Eng. News,* March 6, 1989, p. 14.
4. "Starting Salaries of Chemists and Chemical Engineers—1990," American Chemical Society, Washington, D.C.
5. *Chem. Eng. News,* July 3, 1989, p. 35.
6. Pittsburgh Conference, 1990, unpublished data.
7. "Professional Chemists in Industry," Procter and Gamble Co., Miami Valley Laboratories, Cincinnati.

8. "ACS Cooperative Directory," American Chemical Society, Washington, D.C., 1984.
9. **Laitinen, H.A.,** The role of the analytical laboratory, *Anal. Chem.* **51**:1661 (1979).
10. **Thorpe, T.M.,** Industrial analytical chemistry: the eyes, ears and handmaiden to research and development, *J. Chem. Ed.* **63**:237 (1986).
11. **Half, R.,** *Robert Half on Hiring,* Crown Publishers, New York (1985), p. 169.
12. CFR Title 29, OSHA, Part 1910, "Occupational Exposures to Hazardous Chemicals in Laboratories–Paragraph (f)."

4

Managing Laboratory Personnel

MOTIVATING EMPLOYEES

A question commonly asked is "How can I motivate the people in my laboratory?" Most managers now recognize that the philosophy of "make them do it or else" is obsolete. This philosophy is based on the concept that many, if not most, employees are inherently lazy, do no more than necessary, lack ambition, dislike responsibility, and are indifferent to the welfare of the organization except to the extent that it affects them directly. Laboratory managers who have taken general management courses will recognize this as a description of McGregor's classical Theory X employee.[1] Managers who subscribe to this theory will consider that their responsibility is to get the job done by autocratically attempting to direct, control, and manipulate behavior through a "carrot and stick" approach. The effect is that employees will behave in exactly the way they are perceived and try to get away with as much as possible, without jeopardizing job or earnings.

The progressive view is that managers are responsible for making their subordinates consider themselves as part of the organization so that they, in turn, will feel responsible for helping the organization accomplish its goals. To achieve this objective managers must use their power to see to it that employees achieve the fullest satisfaction in their work, i.e., to help them meet certain psychological or ego needs.

The application of this philosophy is based on a theory that has been described as the hierarchy of human needs.[2] According to this theory, the behavior of people is caused by unsatisfied needs. These begin with the basic requirements of food, clothing, and shelter. Once these physiological needs are met, human beings will develop certain others and will no longer be motivated by what has enabled them to meet the basic needs. New ones arise, the needs related to safety and security. These are followed by certain social needs, and finally by the ego needs. This is not to say that higher needs will be completely absent until the lower ones are satisfied. It is that the lowest unfulfilled needs will dominate behavior and must be addressed in motivating that person.

To translate this hierarchy into terms that relate to the laboratory, the basic need is for an adequate salary. Once this has been attained the employee is more concerned with job security and benefit policies, such as insurance, sick leave, retirement, etc. The next higher need is for social acceptance, the feeling of belonging as part of a group. The ultimate needs are the psychological (ego) ones.

It seems reasonable to assume that in most laboratories today the first three levels of needs are being adequately met. It is the ego needs with which managers must be concerned if their employees are to be motivated to perform at the highest level of their potential. These are

Respect	Appreciation
Status	Recognition
Sense of dignity	Independence
Knowledge	Sense of contribution
Responsibility	Feeling of achievement
Self-confidence	Personal growth

If these needs are not met, employees will feel dissatisfied and perform at the lowest acceptable level; or, they may either resign or approach the manager for what is perceived to be in the manager's power to grant—economic benefits that will enable them to achieve satisfaction outside of the work environment.

In a survey of 1 million workers conducted by the Hay Group Management Consultants in 1989[3] it was found not only that respect was among the highest of desires, but that those who thought they were respected by their boss were three times as likely to be satisfied with their job as those who didn't think so. It may also be of interest to note that when the ACS questioned newly employed chemists on what factors were weighed most heavily in reaching their decision on accepting the offered position, they cited many of the above expectations.

Case History—The Crowded Workbench

Mark, a technician group leader, received a new autotitrator that was intended to speed up a test for determination of neutralization number, a measure of the acid buildup in used lubricants. He approached Ed, the technician who normally ran the test, and said in effect, "Ed, we have this new titrator that you'll have to make room for." Ed bristled. Not only was he not in on the decision to acquire the titrator (unmet ego need), but he was now being asked to consolidate already crowded working space. He objected in no uncertain terms. Mark suspected that the objection was based more on Ed's view of the instrument as a labor-saving device that somehow was a threat to his job than on a lack of space, but he decided to back off temporarily.

Some days later, after Mark discussed the problem with his section leader, Bob, the latter intervened. Bob asked Ed when he might be free to discuss a matter that was giving him some concern. (Note *respect*.) When Ed came into the office, Bob informed him that long-range plans included periodic rotational re-assignments to broaden technicians' experience. (Note *communication* and *personal growth*.) As part of this process he was attempting to make jobs easier for less experienced people to learn (*status* and *knowledge*). At the same time, to keep up with the heavy workload, it was necessary to eliminate some of the routine aspects of the work by introducing automated equipment. He then raised the question of the new titrator. Ed repeated his complaint, but in a more restrained manner. Bob's answer was something along the lines of "Yes, I know you have been trying hard to make the best of a bad situation (*appreciation* and *recognition*). What can we do to relieve it?" "I don't know," Ed replied. The rest of the conversation went about as follows. "Do we have any equipment that is not used very often and is taking up space?" "There might be." "Do you think we could take a good look at the equipment in your area and see what is used least frequently, something that wouldn't be too hard to take down and set up again when needed?" "Yeah, I guess we could do that." "And that would allow us to use the titrator without upsetting things too much?" "Maybe." "Shall we try it?" "O.K."

Case History—The Old Timer

Harold, a 57-year-old technician with nearly 40 years of service, had come up through the ranks. He had worked in the spectrometry group for over 25 years, knew the "ins and outs" of the instruments better than anyone around, and could interpret a spectrum as well as the best. The problem was that he failed to recognize his limitations. He resented having to report to Maury, a relatively new and much younger professional, whose responsibilities included discussing incoming problems with the clients before scheduling work. As a result, Harold grew increasingly surly and become a chronic complainer. His productivity dropped to the point where he did just enough to get by, and he often responded to suggestions with a comment along the lines of "That's no damn good." He would then seem to try to prove he was right, rather than try to make things work.

Maury felt that the situation had reached the point where something had to be done but wasn't quite sure what. After discussing the matter with his supervisor, Maury took the following two ego-boosting motivational actions that were recommended: (1) he frequently had Harold participate in the discussions with clients, occasionally asking his opinion on a detail in the presence of the client, and (2) he asked Harold to serve as an instructor and trouble shooter for other less experienced technicians to be trained in operating the spectrometers. These steps undoubtedly fulfilled emotional needs that had previously been unsatisfied, as Harold's attitude and relationship with Maury improved markedly.

DELEGATION

Delegation is usually viewed as a way to free managers from certain routine aspects of their job so that they will be able to concentrate on their main responsibilities of organizing, planning, coordinating, and controlling. Another consideration is that although a manager may know *what* needs to be done and *why*, a subordinate may be in a better position to know *how* the job should be done. Thus, delegation often serves to fulfill an immediate need in the efficient operation of the laboratory. It is also probably the most effective means of training and developing personnel. Beyond these considerations, delegation is foremost among the means managers have at their disposal to motivate subordinates by meeting their ego needs. Among other effects, it promotes a sense of responsibility and, by indicating the manager's trust, adds to the delegates's feelings of recognition, achievement, and personal growth.

Delegation may be defined as the act of giving a subordinate the responsibility and authority to act for the manager. It will be readily recognized that without the authority the value of the delegation from the standpoint of the delegate will be minimal. It should be emphasized, however, that while the act of delegating may free managers from certain routine aspects of their job and allow them to expand their own activities, it does not relieve them of ultimate accountability. Therefore, the manager must have a reasonable degree of confidence that the delegate can perform the job.

A number of steps may be taken to help ensure that the objectives of the delegation are achieved:

- Select as a delegate one who has performed satisfactorily on an assignment at a level just below that of the delegated one.
- Define clearly the results expected and the delegate's limits of authority.
- Make certain that all persons affected by the delegation are aware of the action and know that the delegate is acting on behalf of the delegator.
- To avoid priority problems and "buck-passing" be careful not to overlap delegated assignments that may have two staff members responsible for the same resources.
- Although procedures may be discussed and suggestions offered, once jobs have been delegated allow delegates to handle them in their own way.
- Arrange for reporting on some agreed-upon basis. Show interest in progress but avoid excessive intrusion.
- Avoid the temptation to go over the delegate's head to correct what seems to be an error or change in plan. Discuss

the matter. The delegate may have had a valid reason for the deviation and, in any case, should be the one to make the correction, especially if other personnel are involved. This will avoid an embarrassment that would completely negate the motivational and developmental benefits of the delegation.

COMMUNICATING WITH EMPLOYEES

The manager's sharing of information and knowledge with subordinates is a form of recognition that is an important factor in determining how employees view themselves and the extent to which they will be motivated. The chemist who has been given a explanation of *why* a job needs doing is far more likely to accord it a higher degree of care than is one who is simply given the task without explanation. Managers who take the time to review the background of a problem and discuss what needs to be done not only ensure that the subordinate has a proper understanding; they also meet many of the ego needs, particularly if the employee's views on alternatives are solicited.

Too often, when an assignment is not completed in the expected manner, the explanation is "I thought that was what you wanted me to do." What causes such failures in communication and how can they be avoided?

Problems arise through the failure of both parties to recognize that for communication to be effective it has to be interactive. To the employee the boss may seem to have expected instant understanding; asking for clarification might be interpreted as confessing to slow-wittedness. Also, intentionally or not, the boss may have created the impression of being too busy, distracted, or self-assured and therefore somewhat unapproachable. The employee will then proceed according to his or her understanding. If several supervisory levels are involved in carrying out the assignment, and communication between them is similarly ineffective, it is obvious that the bench chemist who does the actual work may go far astray from what was originally intended.

To avoid problems of this nature is not at all difficult. It is, of course, necessary to recognize that communication is a two-way affair. What we say is not necessarily what the listener hears. It is essential to avoid the use of terminology that has an ambiguous meaning. Probably every analytical chemist at one time or another has been given a sample with a request for a "complete" analysis. Only after questioning can it be determined how much of an analysis is considered "complete". Not only should sufficient time be allowed for questions, but if they are not forthcoming it should not be assumed that there is perfect understanding. The manager is more likely to identify uncertainty by asking "What questions do you have?" rather than "Are there any questions?". If the discussion has dealt with a complex matter, it is even better to say something along the lines of "To make sure we're both on the same wavelength let's review what is to be done."

Obviously, it is not only in the handling of information relating to work assignments that effective communication is essential. Communication is the means by which the manager controls every aspect of the day-to-day operation of the laboratory. At the same time it is an important factor in helping personnel achieve their ego needs. Successful communication requires one to:

- Listen. All too often in a conversation in the laboratory, as elsewhere, we are so involved in thinking about what we intend to say next that we don't hear what is being said to us.
- Make sure that you understand and are understood. Technical people tend to develop a jargon relating to their specialized activity that is meaningless to the outsider.
- Keep upper management and employees fully informed. Superiors should be advised of employees' concerns and, to avoid unfounded rumors, subordinates should be informed as soon as possible of managerial actions that may affect them.
- Maintain visibility and accessibility. The manager who says "my door is always open" is far less likely to be approached and kept informed than one who is frequently seen in the laboratory visiting subordinates at their work stations. The practice of "management by walking around" has the added advantage of conveying to subordinates a sense of the importance the manager places on their work.

• Personalize announcements. A bulletin board notice may be suitable for communicating certain information, but it is not satisfactory for matters that relate directly to working conditions or the employees' welfare. A personal announcement, made simultaneously to all employees or to key people for simultaneous relay to their own subordinates, allows questions to be asked and avoids the possibility that false rumors will be spread.

A number of suggestions may be offered to help the manager prepare for and conduct discussions. Although some actions are so obviously desirable that mention seems unnecessary, they are often overlooked in the "pressure cooker" atmosphere that prevails in many laboratories.

• Plan ahead as much as possible. An example of the effective use of planning is the preparation public figures often make in advance of scheduled news conferences. By having staff members review the expected questions, the speaker is prepared to handle the responses to best advantage. Likewise, if approached by a subordinate with a complaint, a manager should try to determine the nature of the problem and defer the discussion to allow time for formulating a position.
• Consider the time and place. Some discussions, particularly those that are likely to have a negative impact, have been found to be best held at the end of the work week or before a holiday to allow a "cooling off" period. Others, requiring a reasoned judgment, may not be suitably held at that time when thoughts are apt to be directed toward the forthcoming days away from the job. Similarly, some discussions are best conducted in the privacy of an office; others are better held in the workplace, where they are also less likely to become unduly prolonged.
• Stress the benefit to the listener before the cost. The principle of successful salesmanship can be applied to the selling of an idea. For example, if a distasteful assignment is to be made, the job may seem more palatable to the assignee if developmental, prestige, or similar motivational factors are mentioned first, before the listener reacts negatively.

- Present only one idea at a time. Discussions that cover a range of subjects at the same time are likely to leave participants confused as to which comments related to which topic and what decisions were made.
- Check frequently for understanding. Deferral of questions until the end of the discussion will usually result in some forgotten and unanswered ones. Again, it is desirable to almost obligate the listener to ask questions by soliciting them with words that indicate they are expected.
- When there is a difference of opinion, avoid "yes, but —". To the listener this means "no". Acknowledge the listener's point of view before responding. An immediate defense of one's own viewpoint will only serve to harden the listener's objection. The key to this "sparring" technique is to recognize the listener's ego needs by finding some way of seeming to agree, at the same time formulating questions that will discourage defense of the opposing position; then ask questions pertaining to the objections until the adversarial environment is dissipated. [See "Case History—The Crowded Workbench", this chapter.]
- Observe the listener's facial expressions, tone of voice, and body language. It is usually not difficult to recognize when a point we have made has evoked a negative reaction and will probably be disputed. The interval can be used to plan a response that will best avoid confrontation.
- Avoid trite expressions. Expressions such as "you know", "O.K.?", "see what I mean?", etc. that have tended to become habits of speech can be very distracting to the listener, who will be waiting for the next interjection instead of concentrating on what is being said.

HANDLING PERSONNEL PROBLEMS

The recognition of employees' ego needs in dealing with the daily activities of the laboratory can do much to minimize the personnel problems encountered from time to time. One managerial action that can lead to difficulty is the way in which errors are criticized. Although there is no doubt that serious errors should not be overlooked, the manner in which the manager acts to correct them can have a marked effect on the employee's subsequent behavior. It is essential that any criticism that is made be directed at the act, not at the person. For example, in the case of a test that requires repeating because of some mishap, one might be tempted to ask "What did you do wrong?". The listener hears *you*. On the other hand, if the question is phrased "What do you think might have caused the test to go wrong?", it is perceived as an expression of *respect* for the listener's knowledge.

An even more serious fault in the handling of criticism arises when the manager says, in effect, "You did it all wrong. This is what you should have done." The employee not only takes this as a personal rejection but is subsequently likely to be reluctant to undertake other actions without prior directions. Similarly at fault is the manager who, on being alerted to a problem in a subordinate's area of responsibility, says "Let me think about what we can do" instead of "How do you propose to handle it?". Richard Hagener, a prominent management consultant, has described this as "taking the monkey off someone else's back and putting it on your own". A manager who in this way undertakes to do a subordinate's job is not likely to have time to exercise proper managerial functions.

Although it may be important to criticize errors constructively, many management authorities recommend that greater emphasis be placed on praising and rewarding good performance. This has been called the Theory of Positive Reinforcement. In applying this theory one should look for specific opportunities, particularly after the employee may have been criticized for some failure. To gain maximum benefit, it is also important to minimize the time that elapses between the favorable action and the reward. Rewards need not be tangible; meeting ego needs may be even more effective. These points are illustrated by the following two examples.

Case History—The Temperamental Glassblower
Mike, the laboratory glassblower, had top-notch ability. He could shape a vessel with the skill of an artist. There was one problem. His services were much in demand, and unless he was treated with "kid gloves" one might have to wait weeks to get a job done while others were favored with priority

service. Although strict adherence to order of job receipt was impractical because a simple repair might otherwise be delayed for days while Mike fabricated a bench-scale pilot unit, the analytical people believed and complained that they were at times delayed unreasonably.

One day Mike designed and promptly made a complex distillation column for Art, an analytical group leader, who mentioned it to his laboratory section head. The latter wrote a memo to Mike, thanking him for the excellent job he had done and noting its importance. More significantly, mindful of advice he had once received that it is sometimes more important to consider copy recipients than the addressee, he indicated by an unobstrusive "cc" that a copy was sent to Mike's superior. There were few subsequent complaints about service to the analytical staff.

Case History—A Super Performance

Sam, a senior member of the analytical staff, had just returned from successfully completing a field assignment of major importance to the company. Although there were a few complimentary comments by his immediate supervisor, the achievement seemed to have been taken for granted, despite the fact that Sam had been instrumental in identifying the problem as well as solving it. In fact, it was not until some time later, when a top executive of the field unit wrote a letter commending the laboratory and mentioning Sam's role, that there was any important acknowledgment. The letter was forwarded to Sam by the laboratory director, with the simple comment—"nice work". There was no immediate further recognition, and to Sam's dismay his usual salary review period also passed with no action.

Sam's morale left much to be desired. The effect on his performance must have become evident, since some weeks later the laboratory director took the unprecedented step of inviting Sam into his office. At that time, he mentioned Sam's accomplishment and said that if he was concerned about his annual increase, it had not been overlooked but was delayed for reasons that would soon become evident. Shortly thereafter, the laboratory was reorganized and Sam received a significant promotion along with an appropriate salary increase.

To Sam it seemed that the achievement should have been acknowledged earlier, even without immediate tangible reward.

Relations with subordinates can also be affected adversely by the manner in which the manager responds to memoranda they submit. From the standpoint of the employee, the memorandum may be something that is important enough to warrant early action. The manager, on the other hand, may not be able to make an immediate decision or may put it aside and overlook it for some time, while the employee awaits a response or evidence of action. Although there are undoubtedly occasions when delay is unavoidable or even desirable, some prompt recognition is due the originator of the memorandum. Failure to acknowledge its receipt promptly, with some explanation for any delay or lack of action, is apt to be interpreted as belittling the employee's importance.

MANAGERIAL POWER AND CONTROL

The way in which managers view their responsibility to exercise power and control will have a marked effect on both relations with employees and attainment of the laboratory's objectives. Few managers will succeed if they consider that their position confers on them the right to dominate and *force* desired behavior on their subordinates. Managers are, however, responsible for *influencing* behavior and, as we have seen, can do so by providing the tangible and intangible rewards the employee seeks. They may also influence behavior by virtue of the respect accorded their status, experience, and knowledge. Effective managers use these latter attributes to marshal resources and "cut through red tape", i.e., help subordinates do their job.

Case History—The Dallying Supplier
Tony, who was responsible for operation of the atomic absorption spectrometer, was effectively out of business. He needed a replacement hollow-cathode tube and the supplier's local representative had several times promised it would be sent out the next day but failed to deliver. Tony reported the matter to his superior, Jason, who immediately went into ac-

tion. He bypassed the local office and put in a telephone call to the general sales manager at the supplier's corporate headquarters. He advised the latter that if this was an example of the service provided, the continued use of that company's entire line of instruments in the laboratory might be reconsidered. The needed part was provided within hours.

Managerial control may be viewed in much the same way as quality control of a product, i.e., establishing standards, measuring performance, comparing the two, and instituting corrective action when needed. In the case of day-to-day operations, the objective of an assignment is established, the progress toward its achievement is monitored, and, if necessary, the procedures and resources employed are modified. Control also involves evaluating employees' performance as related to established goals and helping them implement any needed improvement.

To consider first the individual assignment, some managers practice a form of loose control which is actually no control at all. For example, a manager may inform a subordinate of a problem, presumably giving the necessary background information, and saying something along the lines of "Run whatever tests you think and use what help you need, but get the job done in 2 weeks. If you need anything let me know." Although this may foster responsibility and independence, and permit one to learn from mistakes, it is hardly apt to lead to a reasonable rate of personal growth, let alone efficient use of resources. Other managers may tend to the opposite, becoming overly involved and wanting to be informed at almost every step of the way. As with any form of delegation, optimum control will be achieved when there is joint planning and general agreement on a schedule of progress reports, with an understanding that the manager will be informed of any significant deviations from the agreed-upon plan that may become necessary.

Case History—The Perfectionist

Anne was a Research Associate in charge of a small group of chemists responsible for developing analytical methods and handling special problems. George, a Ph.D. from a prestigious midwestern university and a member of the group, was a creative scientist who could work with little technical assistance from others. He could usually be counted on to come up with a suggestion for improving the accuracy or precision of a method or for obtaining additional information on a submitted sample. The problem was that he was too much of a perfectionist and he tended to go off on a tangent. He failed to recognize that the time spent in making the improvement or obtaining more information was often more costly than the value derived. When Anne would query him on when a particular assignment would be completed, his reply was usually along the lines of "I just want to try one more thing." Two, three, or more weeks of effort often resulted.

Anne was reluctant to stifle George, yet the situation was intolerable. After discussing the matter with her section leader she realized that she had been exercising too loose control and that despite George's scientific ability he needed closer supervision. In making or approving subsequent assignments, she specified the goal in no unmistakeable terms. Progress was reviewed on a more frequent basis. When George tended to go beyond the objective, she emphasized the similarity between devoting time to unwarranted research effort and increasing manufacturing costs to purify a product beyond specification requirements. Once, after the objective of a method development project had already been reached, he, as usual, pointed out the scientific value of additional information. This time Anne did not attempt to dissuade him, but agreed to make the facilities available if he wished to pursue the idea on his own time. This, in fact, is what he did, and after some weeks of after-hours and weekend work he completed the project to his own satisfaction.

PERFORMANCE EVALUATIONS

Many managers find that one of the most unpleasant duties they face is the periodic performance review sessions with their subordinates. There is no question of the necessity to evaluate with employees whether goals are being achieved and, if not, what needs to be done to improve their performance and help them develop further. Effective reviews do

just that. Problems arise, however, when the evaluation includes such personal traits as attitude, loyalty, commitment,
cooperation, etc. The views of the employee and the manager
with respect to these characteristics are likely to be quite different, and the discussion may degenerate into a confrontation
or, at the least, lead to a disgruntled employee.

The defects of common appraisal systems were pointed out
by McGregor as early as 1957,[4] and their use of subjective and
judgmental criteria has been criticized by numerous others
since. Difficulties may arise even with well-designed appraisal
systems if, to avoid confrontation, the supervisor is forthright
with respect to positive elements of the employee's performance but is indirect when discussing negative aspects. The
employee is apt to remember only the former and be shocked
when demoted or terminated at some later date for unsatisfactory performance.

In large organizations the laboratory manager rarely has
control over the particular form used for the evaluation. It is
important to recognize, however, that the legal requirements
for an appraisal on which an employee's status is based are
the same as for hiring. To be valid the appraisal must be
consistent with the job requirements as spelled out in the
employee's job description. If the employee's status has been
adversely affected by an unfavorable appraisal that fails to
meet the criteria of validity and consistency, the decision can
be overturned by the courts. The manager must make certain,
therefore, that the employee is evaluated on only those factors
that relate specifically to the job itself.

To deal with the problem of performance evaluations, many
companies have adopted the principle of Management by Objectives.[5-7] This may perhaps best be described as consensus
planning and analysis. There are a number of variations of the
program, but the following steps represent the general form
of its implementation:

1. The employee prepares or updates his or her job description
   and submits it to the manager for review and concurrence.
2. The employee sets targets for those major elements of the
   job to be handled during the established period, usually 6
   months to a year. The goals should be specific in terms of
   results that are quantifiable or can be evaluated objectively.

3. The goals are reviewed with the manager, who agrees if it seems that they are sufficiently demanding yet realistic.
4. At the end of the period the achievements are analyzed and substantiated by statistics, examples, etc., for review with the manager. Any unusual conditions beyond the employee's control that contributed to results significantly different from those expected are noted.
5. Steps necessary for the development of the subordinate are discussed and goals are set for the succeeding period.

The participative and analytical nature of the program, usually referred to as MBO, helps avoid the criticism and negative implications that may be inherent in judgmental evaluation systems. These often create the impression that the manager is like an inspector who always knows best, while the employee is like a product which must meet specifications or be rejected. By being based strictly on performance relative to agreed-upon job requirements and goals, MBO has firm validity under Equal Employment Opportunity regulations.

Some have questioned whether MBO, being based on planned goals, is directly applicable to analytical laboratories, since their activities are often determined by the demands of others. Although it is true that a service group may have less flexibility in planning than other units of the organization, laboratories can in fact meet the criteria of a successful MBO program. Quantifiable goals can be established and achievements evaluated objectively with respect to improving the precision of methods, increasing productivity, reducing turnaround time, implementing new procedures, improving documentation, training for greater flexibility, reducing the cost of supplies, etc.

HANDLING MARGINAL EMPLOYEES

Employees who perform at a minimum acceptable level usually fall into one of three categories: (1) those who are willing but unable, (2) those who are able but apparently unwilling, (3) those who are in a rut. Once the classification of marginal performance has been determined, corrective action becomes more possible. In making assignments, it is well to remember that better people are not necessarily better for the job. In Chapter 1, the general qualifications for the various types of laboratory activity were outlined. With those who are in the first of the above marginal groups, it will usually be possible to find an assignment that they can handle, often in a quite satisfactory manner.

Employees in the second group represent more of a problem and may not respond short of disciplinary action. However, an attempt should first be made to determine why the employee is not performing up to potential. Medical, family, or financial problems may be at the root of the problem. If discussions with the employee aren't productive, a trusted colleague may be able to offer an explanation. Should a definite cause be discovered, the problem may possibly be alleviated by a change of assignment or adjustment of working hours. Help may also be available through the company's medical or employee relations department. If no reasonable cause is found, the appropriate sequence of warnings and disciplinary steps may be the only alternative.

Case History—No Overtime for Me

Ben was a marginal nonexempt employee who had bid on and obtained a "day" job as a laboratory technician, largely on the basis of his seniority. According to the contract with the laboratory union, employees were required to work a reasonable amount of overtime when needed. At the same time, overtime was to be distributed equitably among all technicians qualified to do the work. Ben had always rejected an overtime assignment, usually to his supervisor's gratification, since more productive analysts were generally happy to oblige. (Under terms of the agreement with the union an overtime offer that was refused counted the same as an acceptance in determining equitability.)

One day, Ben was working on a group of analyses that would normally have been completed by about noon the next day. Shortly before quitting time, the client informed the laboratory that it was essential to have the results by morning. On this occasion, others qualified to do the work were absent or otherwise committed. As usual Ben refused to stay overtime to finish the analyses, giving no reason, even after it was pointed out (in the presence of the union steward) that this would result in disciplinary action. With the approval of Employee Relations a 3-day suspension was imposed. Although Ben asked the union to intervene, the penalty was not challenged.

Subsequent discussions suggested that supervision had been lax in allowing Ben to establish a long-time precedent of overtime refusal without trying to determine whether there was some compelling reason. It was also suggested that he should have been offered the alternative of coming in earlier the next day to complete the work in time. As it turned out, however, probably neither step would have changed matters. When Ben left the company shortly thereafter, it was learned that he had been "moonlighting" in another activity that he apparently considered more rewarding; in effect he had been working a 16-hour day.

The third type of marginal performance may be attributable to boredom or, as in the case of the "Old Timer" (p. 48), lack of recognition. Employees in this category can usually be made more productive by a change of assignment or by paying closer attention to meeting their emotional needs.

TRAINING AND CONTINUING EDUCATION

Successful managers make certain that their subordinates develop to the point where they can handle assignments that are progressively more challenging—either in the laboratory or elsewhere in the organization. If not, capable and ambitious subordinates may seek positions elsewhere, leaving the laboratory with a marginally effective staff. This type of stagnation may be avoided by ensuring that at least professionals are given increased responsibilities in progressing from competence in methodology and analytical ability to broader knowledge of the laboratory and company functions. The process may be described as recognizing that there is a difference between 10 years of experience and 1 year of experience repeated ten times.

It is relatively easy to promote the development of technical ability of entry-level professionals. The intensive short courses offered by professional organizations are no doubt familiar to most chemists. These courses provide an opportunity for some to develop a basic familiarity with the subject, while those working in the field may find them helpful in updating their knowledge. Other opportunities are provided through the use of video and computer programs available from professional and commercial sources. Many excellent computer programs, including simulated laboratory operations, are supplied at only a nominal cost by Project Seraphim,[8] inaugurated in 1982 with funding by the National Science Foundation.

Staff members of a laboratory situated near a college or university, particularly in a large urban location where there may be a concentration of such institutions, frequently have available a wide range of part-time or evening courses from which to choose. Tuition fees, and in some cases the cost of books, are generally subsidized by the employer. For professionals at a higher level, the opportunity exists to attend academic departmental seminars; chemistry departments usually welcome participation by people from industry.

Although technicians normally receive most of their training on the job, opportunities of a like nature exist for those seeking to improve their status. It is also possible to obtain college credits toward a chemistry degree from a number of colleges and universities in the U.S. through correspondence,[9] although a period of residence will usually be required before a degree can be awarded. Similar opportunities exist in Canada.

In a less formal manner, technicians who indicate a sincere desire to improve their background without enrolling in an institution may be provided with a technician training text for use in a home study program under guidance of a professional. Several excellent texts are available.[10,11] If several technicians start in such a program at the same time, it may be effective to allot an hour or so during the work week for a classroom style review and clarification of the text material.

An alternative practice in some laboratories is the scheduling of voluntary "brown-bag bull sessions". These consist of a series of lunch-hour meetings at which members of the professional staff each discuss their respective analytical specialities over a period of several weeks. Any staff members who wish to broaden their background in the area under discussion are free to participate.

PROMOTION OF A SUBORDINATE

The use of delegation in connection with development of personnel has already been mentioned (p. 48). In considering which subordinates have what might be called "boss" potential, managers in any field consider such traits as initiative, adaptability, how well they relate to others, their judgment in exercising power, and their urge to assume responsibility. In addition, prospective analytical managers should have at least a general familiarity with the capabilities of the major analytical techniques. They should have shown a tendency to avoid quick, pat answers. For example, the group leader who, when faced with an overload situation, has questioned clients on the need for certain test work, explored the possibility of shifting priorities, or tried to borrow help, is a better prospect than one who immediately requests authorization for use of overtime. The potential manager will also avoid the tendency to get involved in peripheral details. A general inquiry on the status of a sample will not be answered by a detailed review of all the remaining work to be done; nor will a question on the analytical approach being used to solve a problem prompt a step-by-step description of the procedure. Above all, the prospective manager should understand the real role of the analytical laboratory and not have a "black box" mentality (see Chapter 3, p. 34).

Top-level executives often consider that a successful manager is one who has developed one or more ready replacements among subordinate personnel. Indeed, one's own advancement may depend on the ease with which a satisfactory replacement can be found. However, it may not be easy to choose whom to recommend as a successor, or for promotion to a comparable position, if several apparently equally qualified subordinates are available.

An effective way of predicting probable performance on the job is to have had candidates alternate as a temporary replacement in the manager's absence. If the replacement is for a period of more than just a day or two, requiring the acting manager to make certain decisions rather than defer them until the manager returns, a more seasoned judgment is possible. For this reason, taking one's entire vacation in staggered short periods is to be discouraged. Alternating candidates also minimizes the possibility that any one subordinate will be looked upon as the "fair-haired" favorite, with resulting morale problems among others who may consider themselves equally qualified. It should be noted, however, that to foster innovation, upper management will often reject a recommended candidate whom they perceive to be a "clone". Therefore, evidence of independent thought on the part of the nominee is an asset.

As in any appointment or promotion, if and when a selection is made from among several qualified candidates, the others should be informed before any general announcement is made. They should also be assured that they will be given fair consideration for other opportunities that exist or may develop in the future.

REFERENCES

1. **McGregor, D.M.**, *The Human Side of Enterprise*, McGraw-Hill, New York (1960), Chap. 3.
2. **Duncan, W.J.**, *Management*, Random House, New York (1983), p. 263.
3. *Wall Street Journal*, Nov. 17, 1989.
4. **McGregor, D.M.**, An uneasy look at performance appraisal, *Har. Bus. Rev.* 35(3):89–94 (1957).
5. **Klatt, L.A., Murdick, R.G., and Schuster, F.E.**, *Human Resources Management—A Behavioral Systems Approach*, Irwin, Homewood, IL (1978).
6. **Albrecht, K.** *Successful Management By Objectives*, Prentice-Hall, Englewood Cliffs, NJ (1978).
7. **Drucker, P.F.**, What results should you expect—a user's guide to management by objectives, in *Toward the New Economics*, Harper & Row, New York (1981).
8. Project Seraphim, Department of Chemistry, University of Wisconsin, Madison.
9. National University Continuing Education Assn., *The Independent Study Catalog*, Peterson, Princeton, NJ (1989).
10. **Kenkel, J.**, *Analytical Chemistry for Technicians*, Lewis Publishers, Chelsea, MI (1988).
11. **Hajian, H.G. and Pecsok, R.L.**, *Modern Chemical Technology, Vols. 1 and 2*, Prentice-Hall, Englewood Cliffs, NJ (1988–90).

5

Oral and Written Presentations

ORAL PRESENTATIONS

Laboratory managers and their senior aides are often called upon from time to time to make formal presentations relating to their operation. Such presentations may be of a technical nature, as when an analytical research project is being described, or they may deal with administrative matters such as personnel or budgets. A discussion of all of the factors to be considered in planning and presenting a formal talk is beyond the scope of this book; numerous guides are available elsewhere. However, several instances from the author's experience may be worthy of mention.

Consideration of the audience composition can help avoid the embarrassment that may result when differences in background are overlooked. While it is important that one does not talk down to the audience, speakers sometimes tend to overestimate listeners' familiarity with the subject. Almost every field of endeavor has its jargon and colloquialisms, and use of these in a presentation to the uninitiated can lead to problems. If, for example, the audience includes attendees from other disciplines, common laboratory terms and abbreviated forms such as XRF, ICP, TLC, etc. should be defined. At the same time, terms such as ROI for return on investment of DCF for discounted cash flow, which are perfectly clear to financial people, may be meaningless to an analytical group. In preparing for the presentation, the speaker should bear in mind the probable makeup of the audience to be addressed and plan to avoid, or at least explain, terms that may be ambiguous.

Case History—A Question of Temperature

While working on a method development project, Helen, an analytical chemist, had observed that a certain compound showed some unusual catalytic properties. Although the material tended to lose activity with continued use, she found that it could easily be reactivated by simply heating at an elevated temperature. During a project review session attended by members of a group of process development engineers, she reported her findings and suggested that they might wish to explore further application of the catalytic reaction. A somewhat lengthy discussion developed around the question of reactivation, with doubt being expressed by several of the process people that the temperature was high enough to restore the catalytic properties. The question was resolved only after someone realized that analytical people commonly express temperature in terms of degrees Celsius, while engineers used degrees Fahrenheit.

Case History—The 42-Gal Barrel

Dennis was a member of the laboratory staff who frequently addressed social and public interest groups as a volunteer speaker. On one occasion he spoke about the various products derived from crude oil, and described them in terms of gallons of gasoline, heating oil, etc., obtained from a barrel of crude. He then added processing and distribution costs to show that the retail price represented a reasonable profit margin. Some in the audience disputed his figures and complained that the products were overpriced. After some discussion, Dennis realized that the audience had assumed that when a barrel was mentioned, a standard 55-gallon drum was meant. He was able to clarify matters only after explaining that "barrel" is simply the standard term in the petroleum industry to designate a 42-gallon quantity and does not represent an actual container. The problem could have been avoided had Dennis considered that others would probably be unfamiliar with the term.

Anyone who has had occasion to make formal presentations recognizes that an effective way to get the attention of listeners is to indicate at the outset that they will hear something of value to them. Thus, an opening statement that arouses the interest of the audience may be one of the most important elements of the presentation.

Case History—A New Test Request Form

During the time that had elapsed since the laboratory started to update the system of handling test requests, a number of changes had been made. Each time a notice had been sent to the clients, along with a copy of the new test request form, informing them that the change was to take effect on a certain date. This usually resulted in a certain amount of grousing. When a computerized sample receiving, reporting, and archiving system was finally developed, it was found that with each test request additional information, not formerly needed, had to be provided by the requestor. Anticipating some resistance, the laboratory manager called a meeting of the senior project leaders whose groups generated the work requests.

Claude, who was responsible for implementing the system, opened the discussion by informing the group that his purpose was to inform them of a development in the way samples were to be handled that would be sure to please them. He told them that the laboratory had been concerned by the delays in reporting and was now about to install a procedure that would bypass the slow data handling and reporting system, giving them more immediate access to results; moreover, he explained, the new system would ensure greater accuracy in the transmittal of analytical data. Their interest in the development was obvious. He then described the elements of the new system, informing them of the additional information they would have to supply with their test requests. As he had hoped, there were no complaints and their cooperation was assured.

As in any form of communication, speakers should be careful to avoid making exaggerated assertions. Complaints have been registered by clients in the form of "It takes me weeks to get results back" when perhaps just a few samples were delayed. On the other hand, analytical people may make a statement along the lines of "We hardly ever get estimates" when this happens only on occasion. This type of comment can only serve to antagonize the listener and lead to confrontation. The choice of less restrictive wording can usually convey the intended message effectively. For example, *many* can replace *most*, *frequently* may be used instead of *usually*, *sometimes* instead of *often*.

WRITTEN PRESENTATIONS

Many managers face problems in reviewing documents written by subordinates. The difficulty some technical people have with writing has been commented on frequently. In some cases, they may spend almost as much time to prepare a report in a form suitable for distribution to the intended recipients as it took to do the work itself. Merely returning a poorly written manuscript to the original writer for revision often results in no significant improvement. On the other hand, managers should not have to spend time on routine editing of documents when their primary responsibility is to review them for technical content, soundness of conclusions, recommendations, etc. Some laboratories have a competent staff member serve as an in-house editor. Although this may be advisable to ensure adherence to proper format and use of correct grammar, spelling, punctuation, etc., inefficiency results if the editor is not familiar with the subject matter and must consult with the author to clarify meanings or do extensive rewriting.

Numerous publications dealing with the preparation of scientific and technical reports are available, and the concepts they present are applicable to the analytical laboratory as well. Indeed, the principles of clear writing are the same, regardless of the discipline or the subject matter. Observance of a few basic rules, several of which are presented below, will help ensure the readability of one's writing, whether it be in reporting on a research project, describing an analytical procedure, or any other form of laboratory communication. Managers should make certain that subordinates are familiar with these rules and encourage following them to avoid the need for excessive editorial revision.

Consider the Reader

As in the discussion of oral communication, a basic principle is to consider to whom it is addressed. The background and interests of the reader may be such that terms and symbols common to the laboratory might need to be defined. If the results of an analytical study are being reported to a client, it is inappropriate to include the same procedural details that would be included in directions intended for an analytical chemist to follow; if considered necessary, such information may be included as an appendix. Even where a common background exists, it should not be assumed that readers are familiar enough with the particular subject to be able to follow the presentation easily; the writer must make a conscious effort to ensure that they are able to do so.

Avoid Impressive Sounding Words

A problem with some writers is their tendency to use uncommon words which, if not requiring the reader to consult a dictionary, will at least slow down comprehension. Examples of words that have been found in laboratory documents are listed below.

Uncommon word	Preferred word
parameters	limits
redundancy	excess
hierarchical	orderly
convoluted	twisted
adumbration	outline
achromatic	colorless
obfuscating	confusing

Subordinates should be instructed to select words that are in common usage and avoid those that may have obscure meanings.

Eliminate Unnecessary Words

Managers often have to cope with writing that is too wordy for the thought it is intended to convey. The following excerpt from a laboratory report is compared to the edited version: "In order to evaluate the possibility that the sample might contain any components that were amenable to solubilization, a portion of it was transferred to a Soxhlet apparatus and allowed to undergo extraction for a period of 24 hours." With the excess wording eliminated, the passage read: "To see if the sample contained soluble matter, we extracted it for 24 hours in a Soxhlet apparatus." Those who are responsible for preparing documents must be impressed with the need to write as simply and concisely as possible. They should avoid the use of word groupings when a single word will express the same meaning. Some examples are

- In the majority of cases (usually)
- In the event that (if)
- Due to the fact that (because)
- In view of this (therefore)

Minimize Use of the Passive Voice

Scientific and technical writing is intended to convey a sense of objectivity. Consequently, the tendency in the past has been to use the passive voice in reports. This has led to many poorly written documents. Although occasional use of the passive helps avoid monotony, overuse results in awkward phraseology.

Technical style guides generally advise writers to use the active voice in papers submitted for publication. This does not necessarily require authors to use first-person pronouns if they do not wish to do so. The following directions for a laboratory procedure, copied from a published text, are presented in both the original and a revised form.

Into a 100 ml. flask is measured 25 ml. of 0.2 N sodium alcoholate and slightly less than one equivalent of the phenol. The reagent, usually 1.0 g. in amount, is added and the solution allowed to reflux during one hour.

In the active voice, the passage can be worded:

Measure 25 ml. of 0.2 N sodium alcoholate and slightly less than one equivalent of the phenol into a 100 ml. flask. Add the reagent, usually 1.0 g., and reflux for one hour.

The revision is not only less awkward but shorter.

Use Concrete Terms

A common fault with some reports is the use of abstract and indefinite terms that require interpretation or that don't convey the full meaning. Some examples of how meanings have been made clearer are

- "The material bends and decomposes on heating" instead of "The material is subject to thermal deformation and molecular degradation."
- "The solution became cloudy" instead of "The solution deteriorated."
- "Heat the solution to — degrees" instead of "Warm the solution."
- "Add 0.5 to 1.0 g of the reagent" instead of "Add a small amount of the reagent."
- "Allow the reaction to proceed for 5 to 10 minutes" instead of "Allow the reaction to proceed for a few minutes."

Watch Sentence Structure

Many manuals on technical report writing present some basic rules of sentence structure. Nevertheless, a major problem with some writers is the poor way in which their sentences are constructed. Also, sentences are often too long to permit easy reading. One of the causes is overuse of the word "and". This is a perfectly proper connective between two closely related and equally important ideas; however, it is often used to connect two or more thoughts that are best presented as either separate sentences or separated by a semicolon. It is also misused to connect a series of unrelated clauses or two clauses that call for a subordinating connective such as "because" or "if". On the other hand, the author has had to review reports in which the writer has gone to the opposite extreme and used a series of short, choppy sentences throughout the report. Although in those cases meanings may have been perfectly clear, the simple subject-predicate-object form of presentation made for extremely monotonous reading. Managers should impress their report-writing subordinates with the importance of using a balanced choice of sentence length and structure. A combination of simple, compound, and complex sentences of varying length will do much to improve readability.

6

Safety and Environmental Concerns

EMPLOYEE SAFETY AND HEALTH

Laboratory managers have always been responsible for the safe operation of their facility. In recent years, however, they have had to deal with a large number of new regulations concerning workplace safety and health as well as the environment in general. Although a comprehensive treatment of these subjects is beyond the scope of this book, several topics might be discussed briefly.

The manager's role in ensuring that new employees are properly indoctrinated with the importance placed on safe operations has been mentioned (Chapter 3, "Orientation of the New Employee"). In addition, the manager must continually promote safety and be alert to any hazards that develop. Four important steps may be taken to ensure that management fulfills its responsibilities and that the staff remains aware of its concern with safety.

1. First and foremost, set an example. If a manager leaves the office to enter into an area where safety glasses are required, ignoring the rule, even during a momentary visit, is inexcusable.
2. Appoint a laboratory safety committee consisting of employees. Except for a chairperson, who may be appointed for perhaps a year to ensure continuity, the committee members should be rotated regularly so that every employee serves at one time or another. The committee is responsible for ensuring that the laboratory's safety rules

are followed and that good housekeeping is practiced. It makes regular inspection tours of the laboratory, perhaps on a monthly basis, issuing a written report describing any violations or hazards that were observed. Particular attention should be paid to the report of the previous inspection to make certain that any deficiency has been, or is in the process of being, corrected. The committee also investigates all accidents and near-misses and reports to the manager its recommendations on action needed to avoid repetition. It has been found that the change in committee membership for each scheduled inspection helps in several ways. It not only ensures that all employees are kept aware of the rules and are alerted to potential hazards, but it makes it possible for those employees who cited a co-worker one month to be inspected by that same employee the next month. They are thus more likely to make certain that their own operations are free from unsafe conditions.

3. Make certain that someone on the staff is responsible for keeping management informed on regulations issued by OSHA or other safety and health authorities. These include the need to keep employees informed on hazards under the various right-to-know rules and to keep records on exposures. OSHA regulations require that records on exposure to hazardous chemicals be kept for the duration of employment plus 30 years.[1] A listing of chemical hazards is available from the National Institute for Occupational Safety and Health (NIOSH).[2] Among other guides to hazards in the laboratory and means of dealing with them, several have been found particularly useful.[3-5] In case of a specific problem, help is available from the American Chemical Society.[6]

4. Where a pattern of ignoring safety rules develops, be prepared to take drastic disciplinary action. Management can be held liable if injury results, even if it can be shown that the employee's action was in knowing violation of established rules.

Case History—Discarded Safety Glasses

Phil worked in the laboratory as a technician, frequently exposed to the danger of spattering by acid solutions that he and others handled. Although management had emphasized and posted the requirement that personnel in the area had to wear safety eye protection at all times, Phil ignored it. Each time Jerry, his immediate supervisor, called the rule to his attention he would comply; however, after a day or two the glasses would come off again because, as Phil claimed, they gave him a headache.

Phil was a top flight technician. His productivity was above average, and the results of his analyses were consistently of high quality. Jerry was reluctant to discipline him for fear of alienating one of his best employees. Also, he had just been through a grueling union grievance procedure that resulted from a 3-day suspension handed another technician for a different type of infraction and he was in no mood to repeat the experience. Still, aside from the implications if an injury actually occurred, he was concerned that others would follow Phil's example.

Jerry discussed the situation with his section leader, who agreed that after repeated warnings a disciplinary suspension would be justifiable. However, it was decided to try a less drastic step first. The union steward was summoned and the situation was explained in Phil's presence. The steward took the position that the union was equally interested in employee safety, and he warned Phil that if disciplinary action resulted, he could expect no help from the union. The indication that a possible suspension would not be challenged was a form of peer pressure that supported management. Further action was unnecessary—Phil wore his glasses thereafter.

Many laboratories, particularly those engaged in the analysis of organic materials such as petroleum and petrochemicals, are likely to have on hand large quantities of flammable liquids, the escaping vapors of which have been found to be the cause of many industrial laboratory fires. The detection of fire hazards and recommendation of preventive measures are primary responsibilities of the laboratory safety committee. The manager can aid the safety committee in its efforts by stressing the need for a comprehensive survey of the facilities for storing, transferring, and disposal of flammable materials. Details on specific steps to be taken in conducting such a survey and eliminating hazards have been presented by Stevens.[7] The discussion describes what hazards to look for, how such materials should be stored, detection of leakage, spill control, and disposal procedures.

Laboratories often operate on a 24-hour basis, with minimum coverage during the night hours. Although it is customary to have two or more employees present at all times, these employees may not always be in close proximity. Also, there are occasions when a second employee is absent unexpectedly. Thus, the question arises of how to ensure the safety of an analyst working alone in the laboratory. The usual precautions involve notifying security people to check during their rounds, depending on the telephone to call central security or, in some cases, providing the employee with an electronic means of summoning help if needed. These approaches are not entirely satisfactory. Security rounds are usually too infrequent to ensure prompt detection of a problem; at the same time, the employee may be overcome or disoriented and unable to summon help.

To provide for the safety of an employee who may be alone for any length of time, some laboratories use a personal distress device to be worn on the person. Such a device can be activated by a simple flick of the finger to emit a distinctive loud shriek that can be heard at a considerable distance. More important, some models have an optional setting that allows the unit to sound an alarm if the wearer is motionless for more than a few seconds, as might be the case in a sudden collapse. The device is known as a Personal Alert Safety System (PASS), and various models are available from companies that supply fire protection equipment.

WASTE DISPOSAL

The need for appropriate facilities for the disposal of laboratory waste has been noted (Chapter 2, "Physical Facilities"). EPA's regulations under the Resource Conservation and Recovery Act (RCRA), covering the disposal of toxic and hazardous waste from chemical plants, apply equally to laboratories.[8] Where the laboratory is closely associated with a plant, its waste, including surplus chemicals, may perhaps be combined with that of the plant for disposal. It is also possible that the laboratory may qualify under the more liberal "small generator" provisions of the regulations, but the exemptions are so restrictive that few industrial or academic laboratories have been found to qualify.[9]

SOURCES OF SUPPLEMENTARY INFORMATION

Useful information dealing with EPA regulations appears in the publication "Prudent Practices for Disposal of Chemicals from Laboratories".[10] This publication covers all aspects of handling such materials, including identification, storage, recycling, and transportation; it also recommends methods of destruction and disposal.

An additional source of information related to OSHA and EPA regulations is the film library of the National Safety Council.[11] The catalog lists scores of programs for training personnel in hazardous materials management. Programs are provided in either film or video format and are available for rental or purchase.

REFERENCES

1. CFR Title 29, OSHA, Part 1910.20, "Occupational Exposures to Hazardous Chemicals in Laboratories—Access to Medical Records."
2. National Institute for Occupational Safety and Health, *Pocket Guide to Chemical Hazards*, Publ. 85-114, U.S. Department of Health and Human Services.
3. **Bretherick, L., Ed.,** *Hazards in the Chemical Laboratory,* 3rd ed., Royal Society of Chemistry, London (1986).
4. *Prudent Practices for Handling Hazardous Chemicals in Laboratories,* National Research Council, National Academy Press, Washington, D.C. (1981).
5. **Kaufman, J.,** Laboratory safety guidelines, *Am. Lab.* **22**(3):170; (9):6; (11)10, 1990; **23**(9):52, 1991.
6. American Chemical Society, Chemical Health and Safety Referral Service, Washington, D.C.

7. **Stevens, A. M.,** Finding and correcting flammable liquids hazards in the laboratory, *Am. Lab. News Ed.*, June 1990, pp. 32–34.
8. RCRA and Laboratories, Department of Public Affairs, American Chemical Society, Washington, D.C.
9. **Rotenburg, S.L.,** Regional Toxicologist, EPA, Region 3, personal communication.
10. National Research Council, *Prudent Practices for Disposal of Chemicals from Laboratories*, National Academy Press, Washington, D.C. (1983).
11. The Film Library, National Safety Council, 3450 Wilshire Blvd., Los Angeles, CA.

7

Sampling and Sample Handling

Few laboratory managers will disagree with the statement that the greatest problem in providing accurate data is with the samples themselves. Errors that result from nonrepresentative, contaminated, or unstable samples far exceed those caused by test method deficiencies or lack of analytical skill. Unfortunately, the laboratory often has little control over the sampling and is expected to work with what it receives from its clients. Thus, although the analytical data may be correct with respect to the laboratory sample, the client may question its accuracy or, if not, accept it as truly representative of the material of interest and come to a completely erroneous conclusion.

REPRESENTATIVE SAMPLING

Analytical procedures often start with the instruction to take a given quantity of a representative sample. It is clear that the only perfect sample is the entire lot; anything less carries with it a statistical probability that it will be less than representative. Also, the smaller the sample, the greater the likelihood that it will deviate in composition from the whole. Yet this is not always recognized by the client who, in most cases, is the one who has taken the sample. The author has encountered instances when a container of perhaps 250 ml capacity was used to skim a sample from the top of a drum of pellets; this was then submitted for an analysis that was to represent the composition of the drum or even a lot of several drums. At other times, tanks of liquid were sampled similarly or by opening a bottom tap, with the effect of striation being completely ignored. Many managers have no doubt had similar experiences.

In analyzing submitted samples, the laboratory can only ensure that the portion used is representative of what it has received. The client must be made to realize that the quoted precision of a method usually applies only to measurements on samples as received by the laboratory. Overall precision is likely to be poorer, and any deviation from prescribed lot sampling procedures may result in an unacceptable error.

EFFECT OF SAMPLING ERROR ON OVERALL PRECISION

Chapter 10 deals in greater detail with the topic of precision and methods of calculating it. However, an example of the effect of sampling on the precision of the measurement is given below. A more comprehensive treatment and review of the subject has been presented by Taylor and co-workers.[1-3]

The variation in the results of an analytical operation is expressed by the equation

$$S_a^2 = S_s^2 + S_m^2$$

where S_a = standard deviation of results on samples analyzed, S_s = standard deviation of the sampling, and S_m = standard deviation of the measurement on homogeneous material. Ideally, there should be no variation in the sampling and $S_a = S_m$. In practice this is rarely, if ever, the case. If S_s is calculated from the difference, the value can then be used to indicate the extent to which sampling may have to be improved for the precision of the reported results to be acceptably close to that of the measurement itself.

Example

Let S_s and S_m each equal 0.5%. Then

$$S_a^2 = 0.005^2 + 0.005^2 = 25 \times 10^{-6} + 25 \times 10^{-6}$$
$$S_a = \sqrt{50 \times 10^{-6}}$$
$$= 0.0071, \text{ compared with } 0.0050 \text{ for } S_m$$

However, if S_s is reduced to 0.1%,

$$S_a^2 = 0.001^2 + 0.005^2 = 1 \times 10^{-6} + 25 \times 10^{-6}$$
$$S_a = \sqrt{26 \times 10^{-6}}$$
$$= 0.0051, \text{ a negligible difference from } S_m$$

To determine whether samples submitted represent homogeneous material, one may analyze sets of replicates at several different size levels. If the variability between replicates of the same size increases as the size level decreases, it may be assumed that the sample is heterogeneous. This is the basis of the method used by Ingamells and Switzer[4] to determine optimum sample size for the analysis of heterogeneous solids; its effectiveness increases as the concentration of the component of interest decreases.

The Ingamells equation is

$$WR^2 = K_s$$

where W = the weight of sample taken, R = precision (calculated as relative standard deviation), and K_s = a constant. To apply the method, one analyzes multiple equally sized samples of the well-mixed material and calculates the precision (R). K_s is then easily obtained. By substituting the value of K_s and the desired precision for R, the optimum weight of sample to be taken may then be calculated.

SAMPLE CONTAMINATION

Nonrepresentative sampling is by no means the only source of sample-related inaccuracy. Samples may be contaminated prior to analysis in a number of ways, often by the container itself. Although many sampling protocols are quite specific with respect to the type of container in which the sample is to be submitted, most managers will probably agree that samples have been received in almost every type of container imaginable. The author has received samples in beer bottles, food jars, medicine vials, bleach containers, paper bags, etc. It seems reasonable to assume that the chemist who receives a sample

Table 1. Contamination by Sodium in Glass*
(ppm Na found)

	Soft Glass	Borosilicate	Polyethylene
Initial	0.02	0.02	0.02
7 weeks	0.15	0.07	0.01
15 weeks	0.30	0.12	0.02

*0.1 N HCl solution.

submitted with so little apparent thought to its integrity is hardly likely to devote the highest degree of care to the analysis.

Even containers that have been approved for certain tests may not be suitable for other tests to be carried out on the same material. For example, a metal container is generally suitable for most tests on hydrocarbon fuels, but a lead-soldered can is obviously unsatisfactory if a test for trace amounts of lead is to be included.

Other instances of container-caused error can be cited. Glass is not entirely unreactive, and glass bottles may not be suitable for samples to be analyzed for trace quantities of sodium (Table 1). Glass may sorb or desorb other elements as well. Samples blended to contain low amounts of phosphorus have been submitted in glass bottles that had been cleaned with a phosphate detergent and then acid-washed and rinsed; despite scrupulous rinsing, extremely high phosphorus values were obtained. At the other extreme, trace amounts of phosphorus and lead were lost to the surface of new glass bottles that had been acid-washed and rinsed prior to adding the sample. Polyethylene containers, which may be suitable for solid and aqueous samples, are completely unsuitable for hydrocarbons (Table 2).

Related to errors caused by the container itself is the possible reaction of liquid samples with the stopper. For example, samples in containers with cork stoppers may contain substances extracted from the cork or unnoticed small particles of abraded cork that cause high results in analyzing for trace quantities of certain metals. The ash of cork has been found to contain major amounts of Mg, Ca, Ba, and K, lesser amounts of Na, P, Mn, and Fe, and traces of ten other metals.[5]

Contamination before receipt in the laboratory can, of course, occur in many other ways. Analytical chemists are in the best

Table 2. Loss of Hydrocarbons
Stored at Room Temperature
in Polyethylene

	Days stored	Loss (wt%)
Gasoline	2	1.2
	5	2.7
	8	4.8
	12	7.7
	15	10.1
n-Octane	9	6.3
	13	9.9
	38	28.2
	65*	45.3

*After this period of storage, the remaining so-
lution contained 615 ppm of dissolved
polyethylene.

position to recognize those factors that can alter the composition of the sample with respect to the components of interest.
Managers should emphasize the need for laboratory personnel
to work closely with those responsible for sampling to help
ensure that their mutual efforts are not wasted or, what is
worse, lead to false conclusions.

SAMPLE PRESERVATION

Since in many cases some time elapses between sampling
and analysis, precautions may be necessary to maintain sample integrity during the interval. Some materials that are readily oxidized may be stabilized by blanketing with nitrogen or
by the addition of ascorbic acid. Volatile samples can be refrig-
erated or stored in insulated containers packed with dry ice.
Hygroscopic samples may be sealed in containers protected by
polymer film. Other samples may tend to stratify and provision needs to be made for rehomogenizing without difficulty.
A comprehensive description of methods for preserving the
integrity of samples to be analyzed for organic compounds is
given in ASTM D3694.[6] The laboratory should make certain
that clients who do not routinely handle these materials are
informed on how to ensure the stability of any such samples
that they submit for analysis.

Table 3. Determination of
Iron in Used Oil*

	Sample 1		Sample 2	
Run	A	B	C	D
1	58	84	83	82
2	85	123	117	115
3	84	167	163	162
4	128	236	219	226

*Values in parts per million.

For materials that are analyzed on a recurring basis, the sampling protocol will usually describe special steps needed to preserve the integrity of the sample. With nonstandard samples, the possibility that postsampling changes will occur should have been considered in arranging for the analysis. In any case, the laboratory staff should be authorized to reject requests for analyses if samples are not submitted according to the prescribed or agreed-upon procedure.

Case History—Stratification of Iron in Used Oil
A group of four used-oil samples, intended to provide a comparison of engine wear under different conditions, was submitted for the determination of iron content. The iron in such samples is present mostly in the form of suspended matter. The samples were not analyzed until some days had elapsed after they were drawn, and gave the results shown in column A of Table 3. When the client challenged the results the same samples were re-analyzed, this time by a different analyst. The results of the second set of measurements, shown in column B, were in line with expectations.

When the ability of the original analyst was questioned, the cause of the discrepancy was investigated. It was learned that contrary to established practice, which required such samples to be submitted in only partially filled containers so that they could be easily homogenized by shaking before a subsample was drawn, the client had filled the container almost to the brim. The analyst, who was not completely blameless in the matter, had not rejected the samples or tried a different method of homogenizing, but did the best he could with the limited head space. With the second set of analyses, the amount of sample that had been withdrawn for the original tests was sufficient to leave enough head space for suitable homogenization of the iron that had settled out on standing. Accordingly, the analysis was more reliable.

To convince the client that this was indeed what had happened, a new set of samples, properly bottled, was obtained from the original stock and analyzed in duplicate by the first analyst. The results are shown in columns C and D. It will be noted that the results agreed quite well with the results in column B, although they were a little lower. This was to be expected, since the values in column B were obtained on material that had been somewhat concentrated by the removal of nonrepresentative matrix. A material balance on the original samples, based on the volumes involved and the two sets of values reported, gave calculated concentrations in even closer agreement with those obtained on the second lot of samples.

TRANSMITTAL OF SAMPLES TO THE LABORATORY

Samples may be sent to the laboratory through various channels. With samples that are sent from the field, such as those directed to a technical service or independent testing laboratory, there is usually not much choice. They may be sent through the mail, by parcel delivery service, or by messenger; frequently there is little direct personal contact with the originator and limited opportunity to discuss the problem or special needs. In such cases, it is especially important that there be an established protocol for the handling of samples to avoid problems associated with the sampling, container, identification, preservation, etc. Furthermore, with samples that are related to regulatory requirements it is also essential that all of the information needed to establish the chain of custody be recorded. There must also be a clear understanding that if samples are not covered by a protocol they should not be submitted without preliminary discussion with the laboratory.

Samples from on-site operations must, of course, also be submitted according to a protocol, although it is easier to discuss matters with the submitter if there is a problem. Some examples of problems frequently encountered are

• Sample received with no documentation
• Documentation received unaccompanied by sample or incomplete with respect to what is needed
• Insufficient sample for requested tests
• Lack of information on possible interferences
• Estimates not provided

Although it may be necessary in some cases for the analyst to take a firm stand and reject the sample, problems can be minimized by having a central location for receiving and screening samples and test requests (see "Sample Receiving"). It may also help to issue an alphanumeric test index that lists basic information pertaining to the laboratory's roster of standard tests. The index might include the following information:

• Test number
• Test title
• Minimum sample size
• Sampling precautions
• Responsible analytical group
• Special instructions
• Approximate cost

   Samples from on-site sources may be sent to the laboratory in several ways. If frequent sampling and testing is needed for process or product control, and speed is the primary consideration, samples may be sent by pneumatic tube; completed samples and written reports may be returned via the same tube. The advantage other than speed is that operators do not have to leave their post to take samples to the laboratory. There are a number of disadvantages, however. The major one, in addition to high initial cost, is that the system is inflexible. Any change in the location of either the source or the terminal will entail additional expense. Also, the size of the tube may prove unsuitable if operational changes necessitate the use of sample containers of a size other than originally intended. Nevertheless, some plant control laboratories have found the arrangement cost-effective where the nature of the test work is thoroughly standardized and little continuous communication between laboratory and client is needed.

Some laboratories have designated a junior staff member to act as a sample collector for samples that are generated at a number of different on-site locations. The collector makes scheduled rounds of the units, collects the samples, prepares the documentation, and brings them to the laboratory. With reasonably standardized test menus this may have the advantage of overall plant efficiency. If the collector is also responsible for drawing samples there may be an added advantage; the laboratory person may be better qualified than a process operator to ensure that samples are properly taken and their integrity maintained. An off-setting disadvantage is that the procedure is unsuitable if delays of several hours or more in getting the samples to the laboratory cannot be tolerated. Also, management will probably consider the sample collector a part of the laboratory staff, even though the chief beneficiaries are the client units. From the standpoint of the laboratory manager, this may be a serious disadvantage at budget time when "head counts" are made.

In many laboratories samples are commonly delivered by those requesting the tests or by members of their staff. This is not only faster in many cases, but it may provide a better opportunity to raise questions about the sample or the desired analytical work. While such a discussion can prove mutually beneficial, the cost to the client in time value may be high. For that reason, if the client has a subordinate technician staff, one of the latter will often be the one who delivers the samples. This person may be acting only as a messenger and not be in a position to provide any meaningful information. To provide the advantage of improved communication, it would seem advisable that all samples, other than routine ones analyzed by standard tests, be delivered by the client in person or by a knowledgeable staff member who can discuss the problem with the appropriate analytical people. At the very least, the request for analysis should identify the person to whom questions can be addressed.

SAMPLE RECEIVING

Samples are usually received by the laboratory in one of two ways. They may be directed by the submitter to one or more of the test groups that will be responsible for the analytical work or they may go first to a central receiving station. If the sample goes to a test group, that group is responsible for logging in the sample and for routing the sample to others who may be involved in the testing. This mechanism can be used to promote communication between the submitter and the analysts. It also permits the most urgent tests to be scheduled first. However, in the absence of a system to alert other groups that are concerned, the increased queuing time and sample tracking problems may negate the overall advantage. The method probably works best in a small laboratory or where one group performs all of the required tests on any given sample.

The use of a central receiving station overcomes the disadvantages of the above system. Timely, uniform, and complete documentation is more easily ensured, and multigroup assignments may be made simultaneously to minimize queuing time. Less time is required to locate samples, since they may be kept in a central location when not in use for the limited time needed to withdraw any one group's subsample. It does require some procedure to indicate priority in testing where the sample characteristics may change once the container is opened, e.g., if a volatile component is present (cf. Table 2), but this should not be difficult to arrange. The receiving system may be combined with centralized reporting, often as part of a LIMS (Chapter 12). The cost-effectiveness will vary with the size of the laboratory, but it is generally greatest in a large laboratory serving multiple clients.

DISPOSAL OF COMPLETED SAMPLES

Samples on which analytical work has been completed are handled in several ways. From the laboratory's standpoint, probably the simplest arrangement for samples originating on site is to return them to the requestors. This may be because the requestor has asked for the sample to be returned, because it requires special disposal arrangements, or simply because the laboratory lacks room. It is customary, however, to retain the sample for a limited time, depending on space and other considerations, in case the results are challenged.

Occasionally, re-analysis is requested after the sample is returned. If so, clients should know that if the analytical data are in question the same sample must be submitted, even if they prefer to use a new laboratory number to ensure objectivity. Submitters need not inform the laboratory that the sample is one that has been analyzed previously; however, they should be strongly encouraged to disclose the results of the re-analysis to help laboratory management in its quality control efforts (Chapter 9).

Samples from other sources may have to be retained for some time, either because a simple mechanism for returning them is lacking or for other reasons. The retention time will depend on the nature and importance of the samples, their stability, disposal facilities, and the possibility that results will be challenged. A realistic retention time can usually be established in consultation with the respective clients.

Selected samples from among those retained may be re-analyzed for quality control purposes. Some samples of unusual composition or which have been analyzed exhaustively may be kept to be used for reference purposes. Others may be retained and grouped for reclamation; among samples of this type are those that have been assayed for precious metals.

The former common practice of disposing of samples by flushing them down the drain or adding them to solid waste may lead to severe penalties. Unless samples containing toxic or hazardous substances are returned to the originators, the laboratory must dispose of them in accordance with the same regulations that apply to surplus chemicals and other substances covered by the Resource Conservation and Recovery Act.

REPORTING DATA

Requestors of analytical service often ask that results be reported by telephone as soon as available. The practice can lead to problems. Frequently, the client becomes impatient and calls the analyst directly to ask whether the analysis has been completed, interrupting operations. Also, an oral report may be misinterpreted or acted upon before it has been reviewed. Clients should understand clearly that only a written confirmation is official and that any action they take based on other than a documented report is at their own risk.

In a growing number of laboratories problems such as the above are avoided by the use of a computer network or by facsimile reporting. The latter is, of course, more economical and can be particularly useful in a plant where the laboratory is limited to serving a few operating, product blending, or shipping units. It may also be an attractive alternative in more sophisticated applications in that it permits easy transmittal of pictorial data, e.g., chromatograms, IR curves, etc. Electronic mail reporting is an even more advanced form of data transmission that may be attractive where a central computerized facility serves numerous, widely scattered clients.

SAMPLE ACCOUNTABILITY

The importance of proper documentation throughout the entire analytical process will be recognized by any manager who has ever been involved in matters pertaining to patents or to any of the regulated activities. For samples that relate to these, it is essential to adhere to the principle of accountability. This principle encompasses every facet of sample handling from the original sampling of the material in question to the final reporting of results and maintenance of records. It applies equally to the originator and the laboratory, and it comprises what is known as "the chain of custody" as well as all of the documentation related to the analysis itself.

Origination of Sample. In establishing accountability, the person requesting analytical work is responsible for selecting the sample and ensuring that it relates to the problem at hand. The requestor must also record the date and time, and make certain that the sample is in a suitable container, properly labeled and protected. If the sample has been taken by someone else, that individual must be identified. The records must also contain the name of the person who transports the sample to the laboratory.

Receipt of sample. The person in the laboratory who receives the sample must identify it by a laboratory number and date of receipt, with all pertinent information entered into a log book or computer data bank. This information should indicate the receiver and everything needed to relate it unmistakably to the requestor's own sample records.

Analysis. The analyst and method used should be clearly shown in a notebook or on a computer printout. Unless described in detail by the analyst, the method should be one that is readily identified by a reference number. Data must be recorded in ink, with no erasures, and be in a bound notebook or in the form of some other permanent record. Erroneous entries should be crossed out, with the correction indicated and initialed.

Report. The report must include date, sample number, method of identification, and analyst's signature. Any unusual circumstances should also be noted. If the report is submitted via computer, the pertinent information must be available in a form that can be clearly related to the report.

Records. Records relating to regulatory samples or that may be involved in patent action or litigation should be retained for at least 20 years. As noted in Chapter 6, certain regulatory records must be maintained even longer.

REFERENCES

1. **Taylor, J. K.,** *Quality Assurance of Chemical Measurements,* Lewis Publishers, Chelsea, MI (1987), pp. 55–74.
2. **Kratchovil, R. and Taylor, J. K.,** Sampling for chemical analysis, *Anal. Chem.* **53**(8):924A–938A (1981).
3. **Kratchovil, R., Wallace, D., and Taylor, J. K.,** Sampling for chemical analysis, *Anal. Chem.* **56**(5):113R–129R (1984).
4. **Ingamells, C. O. and Switzer, P.,** A proposed sampling constant for use in geochemical analysis, *Talanta* **20**:547–567 (1973).
5. **Milner, O. I.,** *Analysis of Petroleum for Trace Elements* Pergamon Press, Elmsford, NJ (1963), p. 9.
6. ASTM D3694, *Practices for Preparation of Sample Containers and for Preservation of Organic Constituents,* American Society for Testing and Materials, Philadelphia.

8

Workload Management

The flow of work through the laboratory is affected by a number of factors, many of which are beyond the manager's direct control. It is particularly troublesome when others plan a project that will require analytical service without having the laboratory participate at some point. Often, the first indication of a change in the workload is a surge of samples from such a newly initiated project; or a client will require priority service because of some emergency situation. The result is the buildup of a backlog that increases turnaround time, causing general dissatisfaction on the part of clients and a crisis atmosphere in the laboratory.

COOPERATIVE PLANNING

Several steps are suggested to help managers cope with such problems. They can continually stress how test costs may be reduced by having the laboratory participate in the planning. Among such ways is scheduling optimum batch sizes to reduce working time per sample. Also, the pros and cons of shift work and overtime from a cost-effective standpoint can be examined; the improvement in productivity that results from continuity of operations may in many cases more than compensate for the added wage premium (cf. "Handling Overload", this chapter). Additionally, the laboratory may be able to recommend simpler, less costly analytical methods; or, if the objectives of the testing are discussed, more relevant methods may be suggested.

Probably every analytical manager can cite cases where failure to consult with the laboratory in advance resulted in increased costs or the reporting of information that was not relevant to the problem. Two such instances are described below.

Case History—On-Site pH Measurement

Close control of acidity in a process reaction was needed, and the unit operator brought several samples a day to the laboratory to be tested by a pH meter. Each time, the analyst interrupted other work to make the measurement, while the operator stood by waiting for the result. The laboratory proposed that a rugged pH meter be acquired for use at the unit, but the supervising process engineer was reluctant to depend on the operator to use it correctly. When the laboratory continued to question the need for so many measurements, it was learned that control within 0.2 pH unit was acceptable. It was then suggested that narrow-range pH paper would serve the purpose and could be used on site by the operator without difficulty. After a short trial period, during which the operator's results with the paper were shown to be comparable to the instrumental results, the process people agreed to discontinue the laboratory tests.

Case History—Measurement of Vanadate

The operation of a process unit depended on the oxidizing power of vanadate solution. The process people had become accustomed to referring to the active ingredient as "vanadium" and submitted control samples to the laboratory for the determination of vanadium. The laboratory, not having been involved in the planning of the control testing and not being informed on the object of the analysis, analyzed the samples by oxidation-reduction. The method, of course, does not distinguish between the valence forms of the element and total vanadium was reported, leading to the false conclusion that no reagent was being consumed. The analyst could easily have determined the pentavalent vanadium by a simple reductimetric titration had the laboratory been involved in planning the control tests or been consulted beforehand.

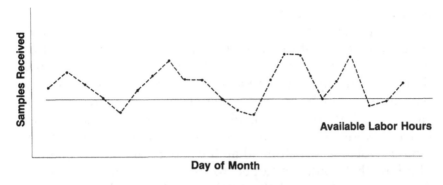

Figure 1. Workload vs. staff.

HANDLING RUSH SAMPLES

Another way in which the manager can help ensure a smooth flow of work is to minimize the impact of "rush" samples. If a client has other samples awaiting analysis, the more urgent samples can be given priority; however, it should be made clear that any samples already on hand from that same client will be delayed to avoid affecting service to others. It may also be possible to have the requestor negotiate with other clients, who may be willing to forego their turn in expectation that the favor will be reciprocated. If the test is a simple one and the number of samples warrants, it may be possible to train someone from the client's staff to perform it; this approach has been found to work well where the complaint is that further activity on a particular project is held up pending receipt of the analytical data. At times it may be necessary to resort to overtime. In that case, the client should be prepared to justify a surcharge (see Chapter 10, "Charging Costs"). Often, when faced with that alternative, the requestor will admit that perhaps a little delay can be tolerated.

OVERLOAD AND UNDERLOAD

If one were to plot the daily workload against available labor hours, the pattern in most laboratories is likely to be as shown in Figure 1. This situation seems to prevail despite improvements in productivity that result from faster methods, since it is the experience of many that the workload always expands proportionately. Although there may be times when the load is less than the staff can handle, the periods of overload predominate. Ideally, the workload and capability of the staff to handle it would be in perfect agreement, with no excess in either direction; this is difficult, if not impossible, to achieve. Indeed, were this to be the case, some higher managements would probably conclude that the laboratory is overstaffed. The best that could be hoped for is that the total area of the peaks does not exceed the total area of the valleys. This requires reducing the peak-to-valley ratio through an increase in the available labor hours, either by staff addition or by other means. Since most managements will approve staff additions only as a last resort, the laboratory must seek other solutions.

Handling Overload

Managers can cope with an overload condition in various ways. As a first step, it is desirable to call the clients' attention to the situation and ask that only immediate needs be considered. They might also be asked to review previously submitted work requests to determine if all still have relevance. In some activities, developments occur so rapidly that test data once thought essential may no longer be needed. Often, raising the question will prompt a client to cancel a forgotten test request of that nature.

Other steps may also be taken. Some laboratories make regular use of parttime and temporary employees. The occasional use of a retired employee for this purpose has been noted (see Chapter 3, "Desirable Personal Characteristics"). If the laboratory is part of a larger technical facility, it may also be possible to borrow help from other activities. In some cases, samples may be sent to another laboratory that is part of the same organization, or to an outside commercial laboratory. If this is done, it is imperative that the laboratory have first established the reliability of that laboratory's data and be prepared to assume full responsibility. Therefore, results should come to the laboratory for review and not go directly to the client.

There is, of course, the use of overtime. In the author's opinion, a reluctance to use overtime because of the wage premium is often unwarranted. When one considers that pay for overtime is calculated on the basis of the employee's *base* salary, the effective hourly cost may be no more than for an equivalent staff addition—possibly less if the cost of all employee benefits is considered. In some cases this added cost may amount to 50% or more.

There are other considerations as well. The occasional addition of an hour or two to a regular work day can usually be counted upon to produce an equivalent amount of productive work. On the other hand, there is no doubt that during the course of an 8-hour day, some time is lost in unproductive activity. A survey reported by Robert Half Management Consultants found that during the average work week in U.S. industry the time lost through waste amounted to 4 hour and 26 minutes.[1]

Another factor to be considered in scheduling overtime to handle an overload situation is the cost effectiveness of continuity. If an extra hour or two will permit the completion of an assignment, it is often preferable to interrupting it to resume work on it the next day. One should be cautioned, however, that continuous use of overtime to keep up with the workload may prove counterproductive and result in little additional output.

Handling Underload

The manager will rarely have a problem in finding ways for the laboratory staff to make effective use of its time in an underload situation. Among activities that will improve overall performance and which might be undertaken or expanded are

- Recalibration and restandardization of instruments and methods
- Training and rotation of personnel
- Preventive maintenance on equipment
- Quality control measurements
- Investigation of new methods
- Housecleaning

Some managers have suggested that clients be informed that the laboratory can handle additional work and be encouraged to submit any samples they have on hand. The consensus among managers with whom the author is acquainted is that this would be unwise, as it would not only result in a flow of perhaps unnecessary work, but might lead to the impression that the laboratory is overstaffed and result in a forced reduction.

ABSENTEEISM

The organization of laboratory personnel in such a way as to minimize the effect of absences has been mentioned (see Chapter 2, p. 18). Nevertheless, the unscheduled absence of a staff member can still disrupt the flow of work and cause a buildup of backlog. Some such absences, whether under an approved quota of allowable "personal leave" days or otherwise, are undoubtedly caused by unexpected personal difficulties that cannot be handled outside of normal working hours.

One factor that needs to be considered is the increase in the number of two-earner families. More and more professional and technically trained women with children under the age of six are working outside the home. Parenting is thus becoming less gender specific. Despite the growing availability of comprehensive childcare facilities, an emergency may arise that will require one parent or the other to be absent unexpectedly.

To enable employees to cope with personal problems and to minimize other problems such as traffic congestion as well, many U.S. companies have adopted the use of nonstandard working hours (flexi-time). The system varies. In some cases, the length of the workday is adjusted so that the days of the workweek are not the same for all employees; in others, the starting time varies, with all employees being required to be on hand during certain core hours. On the basis of the author's discussions with laboratory managers, it is estimated that about 30% use the system in one form or another. Most report that morale has improved and absenteeism has been reduced, but some believe that the ability to exercise proper supervision has been compromised. Others find delays in communication troublesome if not intolerable. It would seem that the nature of the laboratory activities will determine if the advantages outweigh the disadvantages.

USE OF OUTSIDE LABORATORY

In addition to helping reduce backlog in a temporary overload situation, an outside laboratory's services can be useful in other ways. A need may arise for the use of an analytical technique in which the laboratory lacks capability. Even if it is foreseen that the need will recur, it may not be economical to attempt to develop such capability. At times there may be occasion to seek independent confirmation of an analytical result. At other times it may be necessary to have another laboratory serve as a referee in a case of dispute with a customer or supplier. Finally, the objectivity of an outside laboratory may be preferable where a regulatory matter is involved or where the data are to be used in support of advertising claims.

The selection of a suitable outside laboratory is not always a simple matter. There are, however, several directories that can help initiate the process. The American Council of Independent Laboratories (ACIL) has published a directory that provides information on the services of nearly a thousand member firms in the U.S. and several foreign countries.[2] A somewhat larger directory is available from ASTM;[3] this also includes foreign laboratories. Contract analytical services are sometimes provided by large industrial laboratories whose primary function is to support their own organization. Such services should probably be considered only when the use of highly sophisticated analytical technology is needed and the problem is not of a confidential nature.

In evaluating a laboratory that might be called upon for auxilliary service from time to time, its accreditation status should be among the first considerations. If at all possible, the manager or a senior staff member should visit the laboratory to observe its operation and raise certain questions to determine for one's self its suitability. Factors in addition to those dealing with its facilities and personnel qualifications should also be considered. These include quality-control procedures, manner of handling a challenged result, turnaround time, method of reporting, and pricing alternatives.

Among other items to be judged is whether the variety of services offered is consistent with the size and qualifications of the staff. Does the laboratory project a professional image in its documentation, housekeeping, and safety practices? Are its fees unrealistically low? What assurances of confidentiality are offered? Is the laboratory willing to provide the names of others who have used its services? Finally, before a firm decision is made, standard and replicate samples should be sent to assess the laboratory's performance. If the services are subsequently used on a repeat basis, accuracy and precision should be confirmed periodically. A useful guide to evaluating a laboratory is available;[4] many of the above factors are discussed in greater detail.

WORK MANAGEMENT BY COST-BENEFIT ANALYSIS

In deciding to establish or maintain a laboratory, upper management is properly interested only in determining whether the benefits it provides and their value to those who use its services are sufficient to justify the cost. This may be represented by the equation:

$$\text{Value} = \text{Benefits/Cost}$$

The concept may be applied as well to each specific activity of the laboratory. The client who submits a sample or authorizes the laboratory to work on a problem is concerned with the value of the information provided, i.e., do the benefits justify the costs that are incurred. Likewise, as in any business, the aim of the laboratory should be to ensure that it provides its customers with maximum value. This it can do by increasing benefits, decreasing costs, or both.

One basic action that the laboratory can take is to make certain that those who submit samples for analysis are well aware of the costs entailed. Analytical service may sometimes be far more costly than the client expects. The role the laboratory can play in reducing costs by participation in project planning has been noted earlier. The distribution of an index that includes cost information on tests that are run on a frequent basis (see "Transmittal of Samples to the Laboratory," Chapter 7) will also be helpful.

The subject of cost control in the handling of special problems is treated more fully in Chapter 10. However, it is not only the cost of a relatively long analytical study that needs to be discussed before work is undertaken. The advantage of grouping tests to take advantage of the economy of scale should also be discussed with the client. A rapid turnaround may be less important than the reduced cost per sample made possible by withholding the test until a group of similar samples can be analyzed at the same time.

Another way of reducing costs to provide maximum value is to consider the sequence in which tests are run. An order may set up by which the results of one or more tests may be used to determine the need for others on the same sample. Clients will frequently ask for a series of tests on a sample; some tests may be meaningless if a certain value is found on one of those already completed. At other times, analyzing fewer samples than originally intended will provide all of the needed information.

Case History—Accelerated Aging Test

George, a product development engineer, was investigating the stability of a new formulation under accelerated aging conditions. He informed the laboratory that he planned to submit daily samples for analysis to determine the point at which the product began to deteriorate. The project was expected to last for some weeks, but whether the breakpoint would occur in a few days or not for several months was not known. When George was queried on the stability of the formulation at or below room temperature, he was certain that no change could occur during the probable life of the test. It was suggested that instead of having the daily samples analyzed, he store them under refrigeration and have only weekly samples sent to the laboratory. At the point where the weekly sample showed a change from that of the previous week, the intermediate ones could then be analyzed to establish the exact date of break-down. This was done, with the result that the required infor-mation was obtained at little more than one seventh of the cost George would have incurred had the laboratory not ques-tioned the purpose of the test work.

Frequency testing is another method whereby fewer tests can be run, with a statistically determined risk of an unde-tected failure. This is described in detail in Chapter 9, but it involves testing only a fraction of the samples from any given process. The fraction to be tested is based on past experience, the acceptable risk of an undetected failure, and the cost of testing.

The importance of discussing the problem and purpose of the analysis has been emphasized. There is often a consider-able difference between what one asks for and what one really wants; and there may be an even greater difference between what one wants and what is needed. Probably every manager can cite examples of a client asking for some test that was not what was really needed to solve a particular problem.

Case History—Silica/Alumina Ratio

Jack was a young process development engineer experi-menting with silica-alumina matrices intended to provide a base for catalyst formulations. He submitted the first of a group of samples with the request that the percentage of silica be determined. Since such materials were hygroscopic and mois-ture was usually of interest, he was asked whether he didn't really need the moisture to be determined as well; the separate determination of moisture would permit the silica content to be reported on the more meaningful moisture-free basis. He agreed, and modified the request to provide for a moisture measurement.

It was learned subsequently that the moisture was not really of concern in this case. Jack was seeking to control the silica/

alumina ratio and had agreed because he realized that it was needed to allow calculation of the alumina content by difference. He had assumed that a separate determination of alumina would be more costly than a moisture determination. The chemist pointed out that in the absence of other components the ratio could have been determined by X-ray fluorescence directly on the sample as received without the accumulation of errors inherent in the proposed moisture-corrected difference method. This was done with the remaining samples, giving Jack more accurate data at less cost to his project.

Case History—Blocked Turbine Oil Filter

Warren, a technical service engineer, had a problem. A customer was blaming a turbine failure on the oil that had been supplied by the company. It had been found that the failure resulted from a blocked filter that caused oil starvation and the subsequent breakdown. The customer's laboratory had analyzed the deposit that blocked the oil flow and found it to contain a high percentage of zinc, a major component of a metallo-organic compound present in the oil as an antioxidant. The compound was supposed to be stable, but the customer's data indicated that the compound had deteriorated. Consequently, the company was expected to assume responsibility for the damages. Could the laboratory run tests to confirm the composition of the deposit?

When questions revealed the magnitude of the problem, it was suggested that perhaps the laboratory could do more than what was requested and investigate the occurrence more thoroughly. Because of the potential liability, the technical service department readily agreed to fund a more extensive study and several professionals were assigned to the task.

It was eventually found that although the customer's analysis of the deposit was essentially correct, the source of the zinc and ultimate cause of the blockage was not the compound in the oil but the filter unit itself. It had been erroneously constructed of a zinc casting which supported the copper gauze screen that acted as the actual filter. The oil had been contaminated by a leak of waste water containing an ionic compound; in the conducting medium the zinc had deposited on the copper screen through electrolytic action and decreased its porosity to prevent the free flow of oil. When the finding was confirmed by a simulated performance test under the same conditions, the company was absolved of all responsibility. Had the laboratory been content to simply carry out the requested testing, a costly decision would probably have been made.

REFERENCES

1. *Wall Street Journal*, Nov. 4, 1985.
2. "ACIL 1990–1991 Directory—A Guide to the Leading Independent Testing, Research and Inspection Laboratories of America," American Council of Independent Laboratories, Washington, D.C.
3. *ASTM Directory of Testing Laboratories*, American Society for Testing and Materials, Philadelphia.
4. ASTM E548, *Standard Practice for Preparation of Criteria for Use in the Evaluation of Testing Laboratories and Inspection Bodies*, American Society for Testing and Materials, Philadelphia.

9

Quality Performance

The most important responsibility of the manager is to ensure the overall quality of the laboratory's performance. Quality performance comprises many aspects of the laboratory's activities in addition to quality control and assurance, but basically it is the generation of required information as accurately, rapidly, and inexpensively *as warranted by its intended use*. The latter qualification is sometimes disregarded by chemists, whose scientific training has usually stressed the need always to be as accurate as possible.

"Quality giveaway" is a term used in industry to indicate the added cost of producing a material of higher purity than required by specifications. In assessing laboratory performance, it is important to recognize that the term is equally applicable to analytical service. At the same time, in emphasizing speed and cost control, one must not neglect the laboratory's prime responsibility for the production of data that are both valid and reliable. Validity is ensured by the use of methods that measure what, and only what, they purport to measure; reliability is attained when the results yielded by those methods are accurate. Both will be determined by the care with which test methods are selected, how skillfully they are applied, and how thoroughly the results they give are monitored.

CHOOSING TEST METHODS

The choice of methods that meet the above criteria is not always simple. In some cases, numerous methods are available to measure the same property. Many factors may need to be considered in choosing which to use. Some questions to be answered are

- What is the nature of the material or product to be tested?
- What is the authority behind the method?
- What use will the client make of the data?
- Is the method more sophisticated than necessary or will a less costly method do?
- What is the accuracy and precision of the method?
- Are the limitations of the method adequately described?
- Is the client aware of the limitations, the accuracy, and precision?
- Is sampling part of the procedure? If so, who is responsible for it?
- Will the same property be measured at some other point in the handling of the material? If so, by whom and will it be by the same method?

Most consensus methods, such as those of ASTM, AOAC, and other standardizing agencies and trade associations, address many of these points. Other methods are mandated by government regulatory or purchasing agencies, although in some cases it is permissible to substitute a method that can be shown to be equivalent. Useful methods may also be found in published compendia designated "Standard Methods". Although such methods may not have been subjected to the rigorous validation procedures that are followed by standards organizations, they are usually ones that are widely recognized and used; often, however, some modification is needed.

In analyzing materials related to transactions with suppliers or customers, the method to be used is frequently specified in the purchase agreement. If the method is one that is not generally available and has been given to the laboratory by the other party, it will sometimes be on a confidential basis. This requires the laboratory to acknowledge that it will be issued only to those directly involved in testing the material and not released to any others without express authorization.

IN-HOUSE METHODS

At times, a standard method for a particular test or measurement is simply not available. In that case, the laboratory will usually develop its own method or adapt one from a journal or other source to meet the specific need. If intended for repetitive use, such methods will then be adopted as "in-house" methods. To ensure that they are properly documented and contain all of the essential information, they should be prepared according to a prescribed format. Each method should be described by the professional who is responsible for its adoption, and it should preferably be reviewed by a methods editor or documentation group for clarity, conciseness, uniformity of style, etc. The document should include:

- An informative title, a method number, and the date adopted, with reference to previous editions if it is a revision
- A statement on the significance of the method
- The scope of the method and interferences
- Sampling requirements and precautions
- The principle or summary of the method
- Safety precautions
- Needed equipment, reagents, and materials, with suggested source of supply for unusual items
- Directions for calibration or standardization
- Detailed operating procedure
- Method of calculating and reporting results
- A statement on precision and accuracy
- Pertinent references to the source of the method (journal, textbook, internal report, etc.)
- Author and approving authority

Depending on the method, one may want to include such other topics as operating precautions, maintenance hints, reference tables, etc.

BASIC STATISTICAL CONCEPTS

Every analytical chemist should have an understanding of statistical principles to be able to interpret the quality of the data obtained. The following discussion is not intended to be a rigorous presentation but as an introduction to the use of statistics in laboratory work for those managers whose background is in other areas. The subject of statistics as applied to the evaluation of analytical data has been dealt with more comprehensively elsewhere.[1-4]

The validity and reliability of analytical methods are characterized according to the concepts of precision, accuracy, and bias. These may be defined as follows:

• Precision—the degree of agreement of repeated independent measurements with their mean
• Accuracy—the degree of conformity of a value to an accepted or assumed true value
• Bias—a persistent positive or negative deviation of an average value from the accepted or assumed true value

Taylor[1] represents these terms by curves showing the frequency distribution of results obtained in a series of measurements. The sharper the peak, the more precise is the method; the more the peak value deviates from the assumed true value, the more biased (and inaccurate) the method.

The concepts may also be illustrated by Figure 1, in which the center of the target represents the assumed true value of the quality being measured. The plus signs represent individual results that are high and the minus signs those that are low; their distance from the center indicates the magnitude of the error. Figure 1a represents the ideal method—one which is both precise and accurate. Figure 1b illustrates a system which lacks both precision and accuracy; a method giving such a pattern of results would be completely useless. Figure 1c represents a method which is very precise but inaccurate because of bias. Such a method is quite acceptable, however, since once the bias has been established a correction can be applied to all measurements by that method. Indeed, some widely used analytical procedures fit this pattern, with the necessary correction factor being calculated from the results obtained on standard samples.

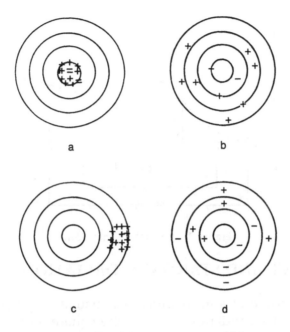

Figure 1. Precision, accuracy, and bias.

Some have expressed accuracy as agreement of the *mean* of a series of measurements with an accepted value. According to that definition, results like those shown in Figure 1d would be considered accurate; the average does not show a bias. From a practical standpoint, however, a method giving such results would have little value because a prohibitively large number of measurements would be needed to establish a mean that is acceptably close to the true value. The figure is included to illustrate that the number of measurements as well as the precision may affect the accuracy of a reported result. In general, however, a method can be precise without being accurate but not accurate without being precise.

One other term that relates to the interpretation of statistical data is "confidence level", the percentage of the measurements that defines the limits of uncertainty. The precision of a measurement is usually quoted in terms of some desired confidence level; in industrial analytical work a 95% confidence level is often considered satisfactory.

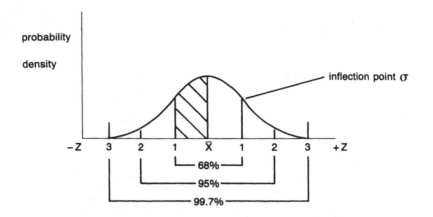

Figure 2. Normal distribution.

VARIANCE AND STANDARD DEVIATION

The precision of a measurement is usually expressed as its variance or its standard deviation, the square root of the variance. If a large number of measurements are made of any given property and the values found are plotted against the frequency with which they occur, a bell-shaped curve will result. The average value, represented by $\overline{X}$, will be at the peak of the curve, the exact dimensions of which will depend on the number of measurements. For an infinite number of measurements, the curve will have the characteristics shown in Figure 2, where σ represents the standard deviation of the measurement.

The area under the curve may be divided into sections, each of which is defined by the value $X \pm Z\sigma$. It will be found, for example, that approximately 68% of the area under the curve lies within the bounds of $X \pm \sigma$ ($Z = 1$), 95% within $X \pm 2\sigma$, and 99.7% within $X \pm 3\sigma$. Thus, if one knew σ and made a single measurement, the value obtained could be stated with 95% confidence to be within 2σ of the average of an infinite number of measurements. Since only a finite number of measurements can be made, the standard deviation of a measuring system is estimated and represented by the symbol S; the value of S approaches σ as the number of measurements used to determine it increases. In this case, the confidence limits are

Table 1. *t* Values at Selected Confidence Levels

D.F.	90%	95%	99%
1	6.31	12.71	63.66
2	2.92	4.30	9.92
5	2.02	2.57	4.03
7	1.89	2.37	3.50
10	1.81	2.23	3.17
20	1.72	2.09	2.85
30	1.70	2.04	2.75
100	1.66	1.98	2.63
∞	1.65	1.96	2.58

given by $\pm tS$, where t is a factor that depends on the number of tests used to determine S.

Some values of t at several confidence levels are shown in Table 1; complete tables may be found in standard texts on statistics. Values are based on the degrees of freedom (D.F.), which is one fewer than the number of measurements made to determine S. From the values in the table it can be seen that in determining S, little will be gained by increasing the number of measurements beyond 20 to 30.

CALCULATION OF STANDARD DEVIATION

By a Series of Measurements on the Same Sample

Record the individual test results and calculate an average.

$$S = \sqrt{V}$$
$$V = (\overline{X} - x_i)^2/n - 1$$

where S = standard deviation, V = variance, x_i = individual test results, $\overline{X}$ = average result, and $n - 1$ = degrees of freedom.

By a Series of Duplicate Measurements

Periodically, as samples are received, split them and enter

them as two separate samples. When 20 to 30 pairs have been analyzed, calculate S by the equation:

$$S = \sqrt{\frac{\Sigma d^2}{2n}}$$

where d = difference between the two measurements, and n = number of pairs tested. In the author's opinion, this method has two advantages over the method based on multiple tests on the same sample. First, all bias is removed if the analysts do not know that they are analyzing replicates. Second, different concentration levels may be included so that S is a truer measure of the overall variation that can be expected. In the latter case, it is assumed that S is relatively independent of concentration over the range tested. This can be determined by the F test (see "Comparison of Precision").

STANDARD DEVIATION OF AN AVERAGE

Intuition tells us that an average of two or more determinations will be more likely to approximate the true value than will a single determination. The standard deviation of an average is expressed as:

$$S_m = \frac{S}{\sqrt{n}}$$

where S_m = standard deviation of the average, S = standard deviation of a single test, and n = number of tests comprising the average. Note that the average of duplicate measurements will improve precision by 41% ($\sqrt{2}$ = 1.41); the average of four will give twice the precision. A decision on whether replicate testing is warranted should be based on the desired precision and the cost of the analysis.

COMPARISON OF AVERAGES

The use of averages will often permit one to determine if two samples that give different values are representative of the same material or not, whereas a single measurement will often fail to do so. For example, a single analysis on two different samples (X and Y) of the same product gives somewhat different results, 10.30 and 10.05. The standard deviation of the method has been established as 0.120 (based on 30 degrees of freedom). The client wants to know if these samples represent the same batch. The value of $\pm tS$ at 95% confidence (see Table 1) is 2.04 $\times$ 0.12 = ± 0.245. With this level of uncertainty for each of the two measurements it is quite possible that the two samples represent the same material. However, if the same values for X and Y were obtained as the average of six measurements each, the confidence limits for each value would be 2.04 $\times$ 0.12/$\sqrt{6}$, or ± 0.100, and the difference would be statistically significant.

In the absence of a known standard deviation it would still be possible to determine whether the two samples differ statistically. This is done by determining a pooled standard deviation (S_p) from the measurements on each as calculated by the equation:

$$S_p = \sqrt{\frac{(m-1)S_x^2 + (n-1)S_y^2}{m+n-2}}$$

where m = number of tests on sample X, n = number of tests on sample Y, S_x = standard deviation of X, S_y = standard deviation of Y, and substituting S_p in the equation:

$$t = \frac{\overline{X} - \overline{Y}}{S_p} \sqrt{\frac{mn}{m+n}}$$

If the calculated value of t exceeds the theoretical value at the desired confidence level, the difference is statistically significant.

Example

Assume that in the above case S_x was found to be 0.14 and S_y to be 0.11. The pooled standard deviation is calculated to be 0.126. Substituting in the equation:

$$t = \frac{0.20}{0.126} \sqrt{\frac{36}{12}}$$

$$= 2.74$$

Since 2.74 exceeds the critical t value of 2.23 for 10 degrees of freedom (95% confidence), it can be concluded that there is only 1 chance in 20 that the difference is statistically insignificant.

COMPARISON OF PRECISION

In choosing between methods an important factor is their relative precision. There may also be other occasions to compare the precision of variables. For example, a manager may wish to determine if one analyst's results are significantly more precise than another's; or, it may be necessary to confirm that the precision established at one concentration level is valid at another. This may be done by means of the F test. The F value is simply the ratio of the two variances (V_1 and V_2), with the higher variance as the numerator of the fraction. This ratio is then referred to a table of critical F values at the desired confidence level and the respective degrees of freedom used to compute the two variances. If the calculated F value is higher than that in the F table, the difference in precision is significant. Table 2 gives some critical F values at the 95% confidence level; as for the t test, more complete tables will be found in standard texts on statistics.

Example

The precision of two methods, A and B, is to be compared. The variance of A is 0.20, based on 8 degrees of freedom; the

Table 2. Critical Values of F at
95% Confidence Level

D.F. Denominator	D.F. Numerator		
	4	8	12
4	6.39	6.04	5.91
8	3.84	3.44	3.28
12	3.26	2.85	2.69
20	2.87	2.45	2.28

variance of B is 0.60, based on 12 degrees of freedom. It is desired to determine whether method A is in fact more precise than method B. Substituting:

$$F = 0.60/0.20 = 3.00$$

From Table 2 the critical value for F at the appropriate degrees of freedom at the 95% confidence level is 3.28, which is not exceeded. Thus, there is a 95% probability that method A is not significantly more precise than method B. If, however, the same variance of A had been based on 12 degrees of freedom instead of 8, the corresponding critical value of 2.69 would have been exceeded, and the method would be judged to be more precise.

QUALITY CONTROL CHARTS

The control of quality by charting is based on several principles:

1. In any measurement system some variation in results can be expected.
2. The variations may be of two types:
   a. Relatively small random and chance variations that are inherent in the system and are difficult if not impossible to identify
   b. Larger variations attributable to some malfunction that needs to be identified and corrected

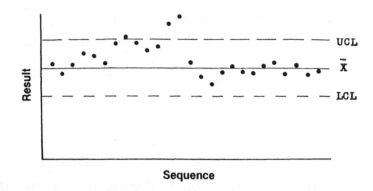

Figure 3. Typical Shewhart Control Chart.

3. Variations of the first kind give a predictable pattern within which statistical laws apply. Variations of the second kind lead to results that exceed the statistically defined limits.

Control charts define acceptable limits to indicate the type of variation so that, if required, corrective action can be taken.

Single Standard Control Chart

This type of chart, commonly known as a Shewhart chart, is illustrated in Figure 3. Values found for successive measurements of the standard sample are plotted on the ordinate and the sequential measurement is plotted on the abscissa. The horizontal line corresponds to $\overline{X}$, the assumed true value of the standard. The parallel dotted lines are at $\overline{X} \pm tS$ and correspond to the upper and lower control limits at the desired confidence level.

In using the chart, one should promptly investigate a point falling beyond the control limits. If it is within statistical probability at the confidence level chosen, additional tests on the standard will give results within the limits. If not, the source of the error should be sought. One should also note if sequential results gradually approach either of the control limits, suggesting a drift out of control.

This type of control chart has the advantage of simplicity. It also permits control limits to be modified if the accumulation of more data allows the value of S to be changed. A disadvantage is that a conclusion on whether the test is in control or not is based on a single standard, which may subconsciously be analyzed with greater care than other samples.

Use of Range to Indicate Variability

The range of two or more measurements may be used to indicate whether a test is in control. To apply the method, the average range of replicate measurements, R, is determined on a series of 15 to 20 samples. (Another method of determining R is to estimate it from a known standard deviation by the use of data in statistical tables.) The value of R is then used to set a control limit for the range of replicates. For duplicate measurements, control limits are

$$2.512 \ R \text{ (for 95\% confidence)}$$
$$3.267 \ R \text{ (for 99.7\% confidence)}$$

Example

Data for the determination of sulfur content had established that the average range of duplicate results was 0.060. The control limit for this test is 0.151 (2.512 × 0.060). As part of the laboratory's quality control program, samples were periodically analyzed in duplicate and the range recorded. The results of these control runs are shown in Table 3. The range of 0.20 in subgroup 12 was beyond the control limit. Accordingly, the operation was reviewed for possible sources of error. After a malfunction was located and corrected, subsequent control runs showed a marked decrease in the difference between duplicates.

The same data are plotted in Figure 4. Plotting is recommended since by presenting the results in chart form one can more easily observe their pattern and be alerted to a possible developing problem. The steep slope between runs 10 and 11 suggests that the test went out of control some time well before it was shown by the results of run 12. Observation of the pattern should have prompted the immediate running of additional duplicates to determine if the value for run 11 was

Table 3. Control Measurements—Range
of Duplicates

Subgroup	Test No. 1	Test No. 2	R
1	1.01	1.05	0.04
2	0.56	0.49	0.07
3	0.87	0.85	0.02
4	0.22	0.24	0.02
5	1.81	1.89	0.08
6	0.98	0.92	0.06
7	1.32	1.25	0.07
8	1.39	1.46	0.07
9	0.15	0.20	0.05
10	1.47	1.53	0.06
11	2.09	1.98	0.11
12	0.70	0.90	0.20
13	1.60	1.62	0.02
14	0.93	0.91	0.02

simply high within the control limits or indicative of a trend out of control.

In evaluating such data, nonrandom variability that should be investigated is suggested not only by a point beyond the control limits, but by:

• One or more points approaching the control limits, indicating a need to accelerate the collection of data points
• A run of three to five points consecutively increasing, suggesting a trend toward the control limit
• A similar run of decreasing points, suggesting that the control limits may be made narrower
• General cycles of drastic fluctuations, which may indicate nonuniform performance by different analysts

INTERLABORATORY TEST PROGRAMS

A major concern in the assurance of quality is how well the laboratory's results agree with those of other laboratories performing the same test. Two methods of evaluating results from laboratories that have participated in a cooperative test program are described on the following pages.

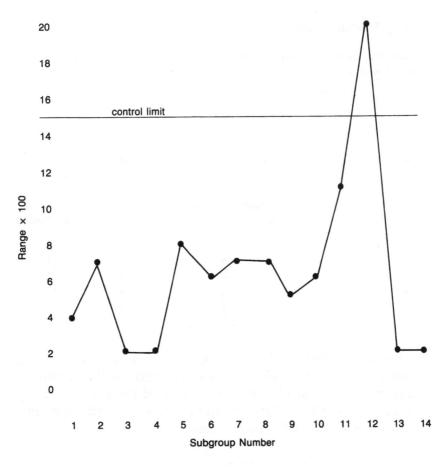

Figure 4. Control chart for determination of sulfur — range of duplicates.

The Youden Ranking Test[5]

It is assumed that m laboratories have tested and reported values for n samples. The laboratories are ranked as follows:

1. List the reported values for each sample in descending order and assign a numerical standing (1, 2, 3, . . . m) to the corresponding laboratory. Thus, the laboratory reporting the highest value will score 1 and the laboratory reporting the lowest will score m.
2. If two or more laboratories report the same value on a sample, prorate the order of standing according to $X + \frac{1}{2}$ for each of two laboratories in Xth place, $X + 1$ for each of three laboratories in Xth place, etc.

Table 4. Youden Ranking Test

Assume five laboratories testing four samples, with the following values being reported:

Lab No.	Sample	A	B	C	D
1		10.1	3.4	6.9	21.3
2		10.3	3.5	6.7	20.9
3		10.5	3.1	6.8	21.0
4		10.0	3.2	6.8	20.8
5		10.5	3.7	6.9	21.5

	Position Ranking				
	A	B	C	D	Summation
1	4	3	$1\frac{1}{2}$	2	$10\frac{1}{2}$
2	3	2	5	4	14
3	$1\frac{1}{2}$	5	$3\frac{1}{2}$	3	13
4	5	4	$3\frac{1}{2}$	5	$17\frac{1}{2}$
5	$1\frac{1}{2}$	1	$1\frac{1}{2}$	1	5

$$n(m + 1)/2 = 4(5 + 1)/2 = 12$$

3. Total each laboratory's score for the n samples. If all of the laboratories have random errors, the scores will be grouped around the average, $n(m + 1)/2$. A total for any laboratory that is high or low relative to the average suggests a systematic error in that laboratory's results.

An example of the use of the test is shown in Table 4. On sample A, laboratories 3 and 5 reported the same highest value; each was therefore assigned a ranking of 1.5. Laboratory 4 reported the lowest result and was assigned a ranking of 5, with the two other laboratories receiving intermediate rankings. The same general procedure was followed for the other samples. In the summation of the rankings, laboratory 5 had the lowest score, indicating a consistent positive bias relative to the others; laboratory 4 scored highest, indicating a negative bias.

Youden has provided a table in which the ranking based on the number of participating laboratories and the number of samples tested are used to identify outlying performance. For the above example the ranking limits are 5 and 19. Thus, no laboratory would be considered to be an outlier, although with laboratory 5 ranked just at the limit the manager would be well advised to look for a possible source of positive bias.

The Two-Standard Test[6]

Although the ranking test is useful in identifying a laboratory that has reported results consistently higher or lower than other laboratories conducting the same test, it gives no indication of the magnitude of the bias. An estimate of how far the laboratory deviates from the mean may be obtained by a second type of program, in which only two samples are distributed to the cooperating laboratories. The samples should be similar in composition and contain about the same amount of the component being measured. The participants analyze each of the samples, making only a single measurement on each.

To analyze the data, a graph is prepared with the scale of the x- and y-axes covering the range of results reported for the two samples. Points are plotted corresponding to the pair of results reported by each laboratory; there will be as many points as reporting laboratories. If the two samples are standards with known concentration of the measured component, these values (X and Y) are plotted and intersects drawn through the point and parallel to the axes. If the exact concentrations are not known, the intersects are drawn in such a way that there are an equal number of points above and below the line parallel to the x-axis and an equal number on either side of the line parallel to the y-axis.

The intersecting lines divide the graph into quadrants. If only random errors are involved, positive and negative ones are equally likely, and the points will be distributed equally between the four quadrants. In fact, experience has shown that most of the points fall in the upper right and lower left quadrants (Figure 5). This is because a laboratory tends to be either high on both samples or low on both. Points in these quadrants but close to the intersection represent results that are only slightly biased and which may well be within the precision limits of the method. The further a point is from the

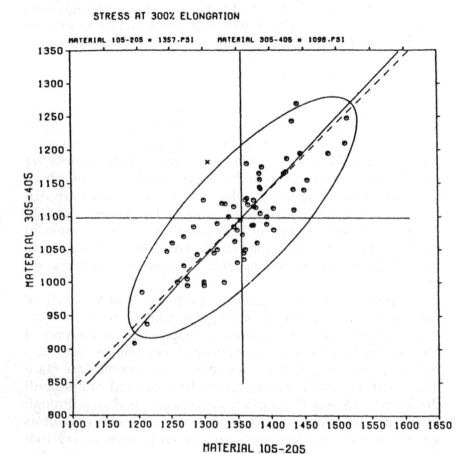

Figure 5. Example of a Youden plot. (Copyright 1974 American Society for Quality Control. Reprinted by permission.)

intersection the greater the laboratory's bias in one or the other of the samples or in both.

If one draws a line at a 45° angle through the intersection, it is possible to obtain an estimate of the laboratory's precision. A point far out from the intersection but close to the line indicates reasonable precision despite the bias. A point away from the line but close to one of the lines parallel to an axis indicates inconsistency in the bias.

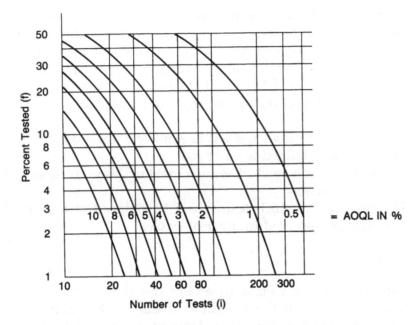

Figure 6. Frequency testing — determination of f, i and AOQL. (From Dodge, H.A., *Ann. Math. Stat.*, 14, 272, 1943. Reprinted by permission of the Institute of Mathematical Statistics.)

FREQUENCY TESTING[7]

A valuable approach to reducing workload from a continuous process or production operation is based on a statistical assessment of past performance. By the method, known as "frequency testing" or "skipped lot testing", a reduced number of samples are tested at a fixed frequency and at a defined percentage of the normal load. The key element in the scheme is the number of successive tests that have shown the process to be in control or the product to meet specifications. The procedure is as follows and is illustrated by Figure 6.

1. Test consecutive samples or batches until a number i have passed without a failure.
2. Reduce the testing to a percentage f of sequential samples from the process stream or product formulation and continue on that schedule until a failure occurs.
3. When a failure occurs repeat steps 1 and 2.

A decision on the amount by which testing can be reduced is based on the risk associated with an undetected failure and the economies that can be achieved. Risk is designated as average outgoing quality limit (AOQL), the maximum percentage of undetected failures that can be tolerated.

In applying the principle one may determine either:

- The number of passes needed to achieve a desired reduction in testing at a defined level of risk
- The risk at some level of reduced testing after an observed number of passes
- The amount by which testing can be reduced at a defined risk after an observed number of passes

Example 1

An operation is willing to accept the risk of passing 5% defective material if the laboratory costs can be reduced by testing only every tenth batch. How many successive passes must be achieved? By following the 5% AOQL curve to the intersection with the 10% f line, we see that 20 passes must be registered.

Example 2

What would be the risk if after 40 successive passes a laboratory reduces its testing of samples from a controlled process to every fourth sample? The intersection of the 40-pass line and the 25% f line is between the 1 and 2% AOQL curves.

Example 3

What percentage of batches must be tested at a risk of no more than 0.5% if the laboratory has observed 100 successive passes? Since the 100-pass line and the 0.5% AOQL curve intersect at the 30% f line, the laboratory will meet the requirements by testing every third batch.

The amount of testing actually done under any of the above schedules will depend on the failure rate, since once a failure is encountered a new series of passes must be registered before reduced testing may be resumed. It is emphasized that any change in the operating conditions at the source of the samples will nullify the procedure and require the accumulation of data on successive passes anew.

REFERENCES

1. **Taylor, J. K.**, *Quality Assurance of Chemical Measurements*, Lewis Publishers, Chelsea, MI (1987).
2. **Youden, W. J.**, *Statistical Methods for Chemists*, John Wiley & Sons, New York (1951).
3. **Miller, J.C. and Miller, J. N.**, *Statistics for Analytical Chemistry*, John Wiley & Sons, New York (1983).
4. **Anderson, R. L.**, *Practical Statistics for Analytical Chemists*, Van Nostrand Reinhold, New York (1987).
5. **Youden, W. J.**, Ranking laboratories by round-robin tests, *Mater. Res. Stand.* **3**(1):9–13 (1963).
6. **Youden, W. J.**, Graphical diagnosis of interlaboratory test results, *Ind. Qual. Control* **15**(11):24–28 (1959).
7. **Dodge, H. A.**, A sampling inspection plan for continuous production, *Ann. Math. Stat.* **14**:264–279 (1943).

10

Budgeting and Cost Control

The laboratory is normally concerned with two budgets—an expense budget and a capital budget. Each is prepared on an annual basis, although the capital budget may include projections for several years in advance to facilitate long-range planning, particularly if the laboratory is part of a large corporate entity. Other things being equal, expenditures under the expense budget are preferable, since costs can be used to offset income for the year in which they are incurred; capital costs must be depreciated according to applicable tax schedules. Under present tax law, however, up to $10,000 of capital expenditure need not be depreciated and may be considered as an expense item.

THE EXPENSE BUDGET

The expense budget consists of the estimated direct labor costs and overhead charges. Overhead includes both indirect labor costs and expenditures for services, supplies, and other items essential to the functions of the laboratory. It also includes a depreciation allowance on the cost of capital equipment purchases.

Typical Indirect Labor Activities
Administration
Vacation, holidays, sickness
Other benefit plans
Other authorized paid absence
Attendance at meetings
Quality assurance
Training
Method development
Maintenance
Sample receiving

Typical Other Overhead Charges
Library, journals
Consumable supplies
Telephone charges
Postage and shipping
Travel expenses
Recruiting
Professional dues
Maintenance and repairs
Rent and utilities
Depreciation
Insurance
Custodial services
General office expenses
Reproduction services

Where the laboratory is part of a larger complex, certain of these other overhead expenses are not easily chargeable to identified users. Accordingly, they are allocated to the laboratory and the other units of the organization. Allocation may be on the basis of staff numbers, space occupied, capital equipment, or other factors judged to be fair by the accounting department.

COST ESTIMATING

The total cost of an analytical operation is usually derived simply from the direct labor cost plus overhead.

$$TC = DL + OH$$

Dividing through by DL we get:

$$TC/DL = 1 + OH/DL = F \text{ (the applied overhead factor)}$$

In calculating the value of F, the previous year's experience of the laboratory is often used. Thus, if labor costs that year amounted to $1 million and overhead $1.25 million, the factor for the current year is 2.25. To calculate estimated costs, one will then normally need to consider only the time to be spent and the hourly wage of the persons who will be assigned to the job.

$$TC = DL \times F$$

Standard Testing Service

The cost estimate of standard tests that are run repeatedly may be expressed as either the average cost per test or the cost per test in batches of n samples. If the average cost is quoted, the cost may be calculated from the total number of tests performed and the total hours expended on them during some selected period (possibly the previous year).

$$(DL \text{ for test/number of tests}) \times F = \text{average cost per test}$$

If the cost per test is quoted on a batch basis:

$$(DL \text{ for batch/}n) \times F = \text{cost per test}$$

where $n =$ number of tests in a batch.

The cost estimated in either of the above ways is not necessarily what is charged to the client. In some laboratory operations, grouping of samples submitted by different clients, dove-tailing of test work, or individual sample variations may determine the time spent on tests in the "standard" category. Charging on the basis of time actually spent is discussed below.

Nonstandard Analytical Service

This type of service may range from relatively simple nonstandard analyses that need only a modest amount of additional effort to major analytical studies that involve committing substantial laboratory resources. Requests of the former type can usually be handled through informal discussion between laboratory personnel and the client. Longer analytical studies, however, should require a more comprehensive discussion and a formal estimate of the cost and time involved.

In furnishing an estimate of the cost of nonstandard service of a project type, one should make certain that the projected time requirement includes more than the actual anticipated laboratory hours. Persons estimating the amount of work needed to complete a task frequently fail to consider the time to be spent in writing reports. Time that will be devoted to literature searches and project-related conferences is also often underestimated.

Although the overhead factor will incorporate the indirect labor and other applicable charges, certain others may need to be included in the estimate. Costs that are directly associated with the project and are not applicable to general analytical activities cannot reasonably be considered overhead. These may include special supplies or reagents, consultation fees, travel, equipment rental, etc.

To avoid any possible misunderstanding, before the assignment is accepted a job initiation form should be filled out. The form should include a number of items in addition to the objectives of the project and a cost estimate. It should indicate an estimated completion date and be signed by the requestor and the analytical staff member making the estimate. Depending on the amount involved, it should be countersigned by the appropriate levels of authority in the laboratory and in the client's organizational unit. If the amount is substantial, the form should also provide a schedule for periodic review of progress. Such a schedule might be based on increments of the estimated cost to ensure that progress is in line with the initial estimate and that the original objectives remain valid. This will not only protect the client against a costly overrun, but it will help the laboratory avoid committing its efforts beyond the point where they are no longer productive.

CHARGING COSTS

Effective control of the laboratory's workload, as well as sound business practice, dictate that there be some system of tracking who uses its services and to what extent. This is done most effectively by an arrangement under which all work that is requested is identified by a job or project number to which the analytical costs are charged. Depending on the organization of which the laboratory is a part, these numbers may be set up to represent specific products, individual processes, different research projects, a special problem, etc. All personnel fill out daily time sheets, relating their time to the appropriate project number or laboratory overhead category. The time is divided into convenient units; increments of 0.1 or 0.25 hour are often used. Non-management supervisory personnel may prorate time not spent working on their own identifiable projects to the job numbers used by their subordinates. These time sheets are then used by the accounting staff to compute the charges against the appropriate client activity.

The laboratory will also have certain charge numbers of its own. In addition to the overhead items, these will include designations for special analytical research activities and overtime that is used for general backlog reduction. Overtime that is used to handle a specific client request may be identified as such when the time is charged against that client's job number, in effect levying a surcharge for rush service.

RELATING COSTS TO BUDGET

In cases where the laboratory's services are used by ongoing or anticipated projects, it is necessary for those projects to include in their own annual expense budgets an allowance for analytical charges. This is usually based on an estimate of the number of analytical hours that will be needed. The cumulative projected total will often exceed the number of labor hours available from the laboratory. Since upper management is unlikely to approve staff additions on the basis of such estimates alone, the discrepancy must be resolved, either by consultation or arbitrary proration of the available hours. The result is that an organizational unit may consume its allotted number of analytical hours well before year-end. Unless other projects have underestimated or can forego anticipated testing, the laboratory is then faced with an overload situation. In fact, this is the situation that often prevails.

It is obvious that the laboratory cannot take the position that since a project's allotted hours have been exceeded no more work will be accepted. Nevertheless, the availability of firm data on how and by whom its time has been utilized will help reduce complaints of delayed service. The information will also serve to justify laboratory cost overruns resulting from increased use of overtime, outside services, temporary help, etc., and support the acquisition of additional resources.

COST CONTROL OF SUPPLIES AND MATERIALS

All but the most modest laboratories will probably have a central source of supplies, although it may be in an attended stockroom if the analytical laboratory is part of a larger laboratory facility. Even if only a small storeroom is used, one person should be made responsible for reviewing and maintaining the inventory of items for general use. All withdrawals should be recorded, either manually or through a computerized inventory maintenance system. It is recommended that no more than a 6- to 12-month supply of any one item be kept on hand, since a change in the nature of the workload may make the item obsolete.

A periodic scheduled "housecleaning" of work areas for the purpose of returning surplus items to central storage is often practiced. This discourages hoarding that leads to duplication and spoilage. If the laboratory is part of a larger installation, a catalog of surplus material may be circulated on a scheduled basis. This listing should be reviewed before purchases are authorized.

In many laboratories the manager has delegated to subordinates the authority to issue purchase orders for expense items; the limit on the amount of money that may be committed is usually based on the delegate's position. The policy not only expedites matters but helps develop the subordinate's sense of responsibility and status as well. Nevertheless, the manager remains responsible for controlling such expenses and must keep informed on all purchases. One way to ensure that this control is maintained is to arrange to receive copies of all purchase orders that have been placed; these may then be reviewed at the manager's convenience.

To reduce the cost of purchased materials laboratories are often able to negotiate substantial discounts from suppliers' catalog prices by agreeing to purchase some minimum amount. Agreements have taken a number of forms. They may provide for a rebate at the end of the year, a credit on purchases once some minimum has been reached, or a reduced price on an initial purchase order for the agreed-upon minimum annual purchase. Another arrangement may simply be an informal verbal agreement to purchase all supplies in a particular category from the same source, with good faith being assumed. In other cases, where the laboratory foresees a continuing need for minimum amounts of certain items, a large initial order may be placed at a discounted price, with the supplier agreeing to maintain the supply; deliveries are then made as needed and billed accordingly. This has the advantage of assuring prompt service and avoids overburdening laboratory storage facilities. Whichever type of agreement is made, significant cost reductions can be achieved in all but perhaps the most modest size laboratories.

EVALUATING STANDARD TEST PRODUCTIVITY

Laboratory managers are frequently required to judge how productivity compares with that of some previous period. Where changes have been made in any one method, a comparison with respect to that method is easy. In many laboratories, however, the workload distribution changes over a period of time. Overall comparisons thus become difficult. If the number of samples is used, how does one allow for the fact that some samples may consume much more time for analysis than others? If the number of individual tests is used, how does one handle the situation in which multiple components are often determined in a single operation, e.g., by gas chromatography or by atomic absorption, whereas in other procedures that are equally or more time-consuming only one component is determined? The time used to calculate average cost per test as described above provides a means of equalizing the discrepancy.

Example

Divide time into designated test cost units (TCUs). The size of a unit depends on the overall nature of the test load; where the test roster includes a number of short tests it may be convenient to have units of no more than 10 or 15 min. To each test assign a TCU value representing the average time required. Where two or more components may be measured in the same processed sample, the TCUs for each will represent the total TCUs divided by the number of components.

For example, if the unit selected is 10 min and the average time for analyzing a sample for five metals by atomic absorption is 30 min, the TCU for each metal is 0.6.

In applying the method to measure productivity, establish a base TCU value for each standard test in the roster and multiply by the total number during a given period to give the overall TCUs produced by the laboratory. The value for each test is maintained and applied to all subsequent tests for the same component or property. As methods or operations are improved, the decrease in time required will be translated into an increase in TCUs that can be produced by the laboratory staff. Thus, if the atomic absorption method mentioned above is extended by 5 min to include an additional element that formerly had a TCU value of 3, a gain of 2.5 TCUs is realized. The laboratory's greater output of TCUs resulting from an increased workload or the addition of new tests can then be related to the base value to give a measure of the progress at any given time.

11

Capital Investments

The acquisition of a major item of capital equipment involves both technical and financial considerations. Decisions on such purchases are usually based on achieving one or more of the objectives discussed below. In essence, they represent monetary savings—directly or indirectly. Where such savings can be demonstrated, the laboratory will be in a better position to compete with revenue-producing units of the organization for upper management's budgeting of funds and authorization for the expenditure.

A primary justification for a large capital purchase by the laboratory is the reduction in operating costs that it will allow. This may result, for example, from installing an alternate means of measurement that provides savings in the cost of labor and/or supplies. Often, too, with an instrument that has been in use for some years, one encounters a prohibitive number of breakdowns that cause intolerable costly delays and/or expensive service calls; in some cases, needed parts may no longer be available and replacement then becomes unavoidable.

Another justification may be in the ability to improve the quality of analytical services—by providing better accuracy, more precision, or greater sensitivity or range of applicability. Also, it may be possible to reduce turnaround time which, in many operations, is of critical importance. Since these steps benefit the client groups more than the laboratory, they can usually be translated into overall savings for the organization. Therefore, if an instrument can be shown to provide improvement in these respects, the support of client groups can often be obtained to help gain approval for the purchase. Indeed, the benefit to the clients can at times be so attractive that they may agree to transfer part of their own capital appropriation to the laboratory.

Two other types of capital expenditures can be expected to receive prompt and favorable consideration by management. One is for any item that improves safety, particularly if it involves an OSHA requirement; the other relates to new test requirements that may be mandated by the EPA or other regulatory agencies. Although the value of such expenditures can hardly be expressed in monetary terms, management will generally recognize that they represent an unavoidable cost of remaining in business and approve the acquisition.

RENTAL OF CAPITAL EQUIPMENT

Prior to purchasing a capital instrument, or perhaps as an alternative, it may be desirable to consider rental. This is often preferable if a continuing need for the particular application is not certain and other ways of obtaining the information, for example, by use of outside service, are not available. If the need is immediate, rental will usually provide an instrument sooner than purchase; lead time on delivery of new instruments may be several months. In fact, rental may be the only alternative if an unforeseen need arises and capital funds are not available or if they can be used more productively for other purposes.

Rental can offer a number of advantages in other respects. It permits the trial, perhaps of several competitive instruments, before a decision is made on the purchase. Lease agreements frequently include an option to buy; the first-hand experience can help avoid a costly mistake. Rental also provides protection against obsolescence; instrument developments are often so rapid that new models may be available before the new instrument warrantee expires. Also, not only is service on a leased instrument usually more prompt, but low-cost maintenance can often be provided because of the lessor's experience with the equipment.

There are, of course, some disadvantages to renting instead of purchasing outright. The most important is that, depending on the term of the lease, overall costs can be much higher. Also, acquiring the equipment free and clear when funds are available avoids saddling the laboratory with a fixed expense for the entire term of the lease. Another disadvantage is that any modification of leased equipment is not permitted and relocation without the lessor's permission may be prohibited.

One factor to be considered in deciding whether to rent or buy outright is the tax consequences. Rental costs are considered expense items and, as noted (Chapter 10, "The Expense Budget"), are deductible from income in the year in which they are incurred. The value of the deduction may far outweigh the depreciation allowance on the item for the year in question. On the other hand, leasing may result in the loss of a possible investment tax credit. Also, if it appears that the equipment was leased for the sole purpose of accelerating tax deductions, particularly if it is subsequently purchased, the transaction may be ruled a conditional sale and the expense deduction disallowed. Where a choice exists, it would be well to review current tax considerations with the accounting department before reaching a decision.

SELECTION OF CAPITAL EQUIPMENT

Ensuring that a particular instrument or piece of equipment will perform as intended is not always a simple matter. Applications in different laboratories may vary, so that satisfactory performance in one is no guarantee of the same in another. With all but standard items it is essential to be as specific as possible about the intended use. Representations by sales personnel is certainly no assurance of satisfaction. To help make certain that the item will do the job, a number of steps may be taken.

- Send samples to the manufacturer's applications laboratory and, if possible, arrange to have someone present when they are analyzed. Even this may not be sufficient; instruments have occasionally been found to be so sensitive that they perform satisfactorily in one location but fail to do so in another.

- Obtain a list of other users from the salesman, and ask them about their experience. This, too, may not be entirely reliable; the list will probably not contain the names of customers who have complained. Professional contacts are a much better source of information. One group whose purpose includes exchange of experience with the performance of instruments is the Analytical Laboratory Managers Association (ALMA); membership is open to all laboratory supervisors and managers.*
- Look into the service record of the manufacturer. Experience with other instruments on hand from the same manufacturer should give a clue, but the particular instrument line may be a factor. Again, consult other users, keeping in mind that the geographic location of the laboratory is often a consideration.
- Unless the item requires significant installation effort, try to obtain it on loan for trial. This sometimes requires submitting an approved purchase order in which the trial period is expressed as a time limitation for return; in other cases, the trial may require payment of a handling fee if the purchase isn't made.

In addition to the above precautions, several other steps have been found to help reduce costs and avoid possible problems.

- Try to obtain modular equipment that will permit selection of only needed options and that can be expanded as additional needs develop.
- Consider used, reconditioned, or demonstration models of proven performance provided an adequate warrantee is offered. Demonstrators, with new instrument warrantee, may be obtained at substantial discount from list price.

*Analytical Laboratory Managers Association (ALMA), P.O. Box 258, Montchanin, DE 19710.

- On other than off-the-shelf items, get guaranteed delivery and installation date and, if the nature of the equipment warrants, a performance guarantee. Include a clause that provides an appropriate penalty if the terms of the guarantee are not met.
- Be especially wary of equipment from foreign manufacturers unless they have been proved to have adequate domestic representation for parts and service.

Where a new instrument is being considered for replacement of one that has had numerous breakdowns or is technically obsolete, it is common practice to seek a trade-in allowance. To retain customers, manufacturers will frequently grant one even though from their standpoint the instrument may be worthless. Some laboratories have found that where a competitive instrument is equally suitable, a more generous allowance will be granted by the competitor. If, however, the instrument to be replaced is one of a line on hand from the same manufacturer, the value of the greater allowance should be balanced against possibly complicating service problems by the need to deal with two different suppliers. The possible need for additional training on a different instrument must also be considered.

SERVICE ON CAPITAL EQUIPMENT

The cost of maintenance, including internal servicing, is high, reportedly up to 30% of the cost of the equipment in some instances. A decision on whether to have instruments covered by a service contract or to operate on a call-in basis is often difficult. Factors that need to be considered include the number and type of instruments from the same manufacturer, in-house capability, and location of the laboratory. A laboratory situated outside of an urban area is likely to encounter intolerable delay in service if it operates on a call-in basis; in that case a contract, including a specified response time, is almost essential. On the other hand, a laboratory close to the supplier's facility can usually depend on prompt service, whether under a contract or not. Many managers have found, however, that with the increasing complexity of instrumentation, a service contract is preferable. Terms are usually negotiable, but for a contract covering parts and labor for a single major instrument, to take effect at the end of the warrantee period, an initial fee of 10% of the original cost is common.

ESTIMATING VALUE OF AN INVESTMENT

In seeking approval for a capital expenditure it is usually not enough to indicate that the investment will result in a given level of savings. It is necessary to show how the conclusion was reached and what alternatives were considered. To communicate effectively with upper management on financial matters, the laboratory manager should be familiar with some basic accounting principles.

Investment decisions are based on the amount of either income or cash flow that the expenditure will generate. Cash flow is the sum of income (after tax for a profit center) plus depreciation. In illustrating the application of the principles to laboratory acquisitions, net savings will be used instead of income. Thus, to simplify, if a particular $20,000 investment is projected to produce savings of $2000 every year, and it is depreciated at the rate of $4000 a year, the annual cash flow over a 5-year period is $6000.

Cash flow and savings are used in several ways to calculate the estimated value of an investment.

Investment on Basis of Payout Time

Payout time is calculated by dividing the cost by the annual cash flow. If the amounts mentioned above are taken, the payout time is $20,000/$6000 = 3.3 years. A short payout time is attractive since it minimizes the risk that the expected results won't be realized; it also frees funds for reinvestment sooner. Where a choice is to be made between purchases, however, one with a shorter payout time is not necessarily preferable. One reason is that the payout method fails to consider possible returns after the cost has been recovered. Thus, one instrument may be obsolete at the end of the payout period, whereas another may continue to yield savings for years afterward. Also, as shown in Example 2 (p. 145) the rate at which the payout time is reached is a factor.

Return on Average Investment (ROI)

In any investment, one is concerned with the return that may be expected. To calculate the return on a capital purchase, it is first necessary to determine the average investment (AI).

$$AI = (cost + \Sigma \text{ undepreciated balance})/\text{years to depreciate}$$

Depreciation is discussed in somewhat greater detail below, but to use the same $20,000 investment and depreciation rate cited above:

$$AI = (20,000 + 16,000 + 12,000 + 8,000 + 4,000)/5 = 12,000$$

It will be seen that for a straight line depreciation (cost/years), the same result is obtained more simply by:

$$AI = (cost + \text{final year undepreciated balance})/2$$

The return is simply the savings divided by the average investment. In this case, the savings of $2000 a year divided by $12,000 gives an annual return of 16.7% over a 5-year period. This level of return may or may not be acceptable to upper management. Factors that will be considered include alternative uses for available funds, prevailing interest rates, and the degree of risk. If the cost of borrowing funds is high, a higher ROI will be demanded than when interest rates are low. To compensate for risk, some managements seek a return of at least twice the prevailing prime rate.

Discounted Cash Flow (DCF) or Net Present Value (NPV)

The basis of this method is the present value of money to be received at a future date. Present value may be looked upon as the reverse of appreciation in the value of an investment that results from inflation, the compounding of interest, etc. The principle may be illustrated as follows: If one assumes that funds are invested at an interest rate of 10%, each $100 will obviously be worth $110 after 1 year, $121 after 2 years, and so on. By the same token, if one receives $100 after 1 year, assuming the same rate of interest, about $91 must have been invested. The same amount received in 2 years will result from an investment of about $83, in 3 years, from an initial $75, etc. The $100 that is received in this way is said to have a discounted cash flow of 10% over the indicated number of years.

The same reasoning can be applied to other discounted percentages. A table covering a selected range of values is shown in the Appendix.

In applying the concept to decisions on acquiring laboratory equipment, one calculates what the return is likely to be in forward years and the present value of each year's return at a selected discounted percentage. The return, of course, need not be uniform; indeed, it usually is not, possibly because the workload changes or the instrument requires some time before it is completely functional. If the benefits provided by the present value of the total return are sufficient to compensate for the risk, then the DCF can be considered favorable; or, if a choice is to be made between two or more instruments that are otherwise equally desirable, the one that will yield a return with the greatest present value would be preferable. For laboratory apparatus a 5-year DCF calculation is often used; this coincides with the depreciation period prescribed for that class of equipment under present U.S. tax regulations.*

Example 1

A return of the cost based on a 5-year DCF of 15% is required on the purchase of a spectrometer for $20,000. Would this be realized if we assume the same $6000 annual cash flow that was used in the illustration of the payout time and ROI methods?

*To illustrate principles the 5-year depreciation rate used in examples has been somewhat simplified.

Table 1. Calculation of Return Based on Discounted Cash Flow

Years Hence	Estimated Cash Flow	Factor	Present Value
1	$6,000	0.8696	$5,218
2	6,000	0.7561	4,537
3	6,000	0.6575	3,945
4	6,000	0.5718	3,431
5	6,000	0.4972	2,983
Present value of return			20,114

By use of the 15% discount table in the Appendix, we obtain the results shown in Table 1.

Example 2

$30,000 is available to be invested in needed new equipment. A choice is to be made between two instruments, both of which will be immediately helpful although used for different purposes. Each is expected to have a 3-year payout. Instrument A will probably be obsolete in 3 years, while B is likely to have a useful life of only 2 years; both will then contribute only $6000 annually in depreciation to cash flow. Instrument A will yield savings of $4000 in test costs during each of the 3 years it is in use, whereas B will yield an initial savings of $10,000 but only $2000 in the second year. In view of the expected early obsolescence, management requires ultimate recovery of the cost on a 15% DCF basis. Which instrument is preferable, other considerations being equal? The comparison is shown in Table 2.

Use of the DCF method with ROI will also sometimes lead to a different conclusion than ROI alone.

Example 3

A choice is to be made between two equally desirable instrument purchases. One instrument (A), costing $20,000, will save no money the first year, but with increased use is expected to save $1000 in the second year, $3000 in the third, $4000 in the fourth, and $5000 in the fifth. The other instrument (B), costing $15,000, is expected to allow immediate sav-

Table 2. Comparison of Payout Time and Discounted Cash Flow

		Cash Flow	
	Year	A	B
	1	$10,000	$16,000
	2	10,000	8,000
	3	10,000	6,000
		30,000	30,000
	4	6,000	6,000
	5	6,000	6,000

		Discounted Cash Flow	
Factor	Year	A	B
0.8696	1	8,696	13,914
0.7561	2	7,561	6,049
0.6575	3	6,575	3,945
0.5718	4	3,431	3,431
0.4972	5	2,983	2,983
		29,246	30,322

ings of $1800 a year and continue at the same rate for the 5-year depreciation period, after which both instruments will be obsolete. Management requires a minimum 20% return on investment. Which purchase is financially more attractive?

Table 3 shows the computation of the ROI based on the savings and depreciation schedule. Table 4 shows the cash flow and the total recovery by the DCF method. The calculation in Table 3 suggests that the return on A is better than that on B. However, when the present value of the return is considered (Table 4), B returns a higher percentage of the invested funds.

It is apparent from these examples that it is beneficial to achieve maximum early return. One way that this can be done, other than in the use of the equipment itself, is by accelerating the depreciation. As noted, current U.S. tax law fixes the depreciable life of laboratory equipment at 5 years, but the depreciation rate is not quite straight line as used in the above simplified illustration of principles. Also, the accounting department would most likely use a permissible nonuniform depreciation rate that yields a more favorable early return.

Table 3. Comparative Returns on Average Investment

	A	B
Cost	$20,000	$15,000
Annual depreciation	4,000	3,000
Average investment	12,000	9,000
Total savings	13,000	9,000
Average annual savings	2,600	1,800
Return on average investment	2,600/12,000	1,800/9,000
	= 21.7%	= 20.0%

Table 4. Comparative Recovery by Discounted Cash Flow

	Cash Flow			Present Value	
Year	A	B	Factor	A	B
1	$4,000	$4,800	0.8333	$3,333	$4,000
2	5,000	4,800	0.6944	3,472	3,333
3	7,000	4,800	0.5787	4,051	2,778
4	8,000	4,800	0.4823	3,858	2,315
5	9,000	4,800	0.4019	3,617	1,929
			Total	18,331	14,355

Recovery A = 18,331/20,000   Recovery B = 14,355/15,000
            = 91.7%   = 95.7%

Another factor influencing return is the time of the year at which the purchase is made; a year-end purchase offers certain advantages from a cost-recovery standpoint. A discussion of this and other matters related to tax law is beyond the scope of this book, but it will suffice to point out, for example, that the manager contemplating a purchase in January or February is often well advised to complete it in December of the preceding year if funds can be made available.

For the above reasons, when an investment decision is to be made, the financial ramifications should be reviewed with the accounting people. This will often help the manager reach the most advantageous conclusion and make it easier to obtain higher management's approval for the purchase.

12

Information Management

Probably no development has had as great an impact on the management of the analytical laboratory during the past decade as that of laboratory information management systems (LIMS). The proliferation of publications on the subject since the definitive articles by Dessy,[1] and the number of vendors of such systems, attest to the significant contribution they are making to improving the efficiency of laboratory operations. It would be well, however, to define what is meant by the term.

WHAT IS A LIMS?

In its simplest form a LIMS is a computer-based manual system at a single work station by means of which samples are logged in with the appropriate identifying information, test assignments made, analytical results entered, and a report generated to be mailed back to the client. This type of system may be operated with only a single personal computer.

More sophisticated multicomponent versions incorporate verification and approval mechanisms, produce reports for management on work in progress, backlog, productivity, and utilization of resources. They also provide an audit trail, compute and assign costs, accumulate archival information, etc. Bar coding is often added to identify particulars relating to source, generate a test menu, facilitate sample tracking, and designate specification limits. One or more terminals are conveniently located for the entry of test results by the analysts from a notebook or instrument readout and for the retrieval of

149

stored information. Access codes may be provided to ensure the integrity of the system within the laboratory.

Essentially all systems now in use incorporate most of the additional features described above. Many, if not most, also include certain others to provide for more rapid capture and analysis of data, documentation, and storage and dissemination of information. These systems combine the data handling features of the laboratory's instruments and computer facilities with the information processing capabilities of the LIMS by such improvements as:

• Interfacing instruments such as spectrometers and chromatographs to allow direct data entry without intervention by the analyst
• Including statistical quality and process control data and trend analysis
• Producing spreadsheet and graphical presentations

Many LIMS installations are networked to computers used by the clients to permit immediate access to analytical data. This is particularly valuable where delay can prove costly as in the case of process control, product formulation, and shipping decisions. A LIMS may also be used as an electronic methods manual, simplifying the maintenance of up-to-date and uniform laboratory procedures, and as a means of tracking inventory of supplies and materials. Advances are also being made in incorporating decision-making features, such as modification of test parameters, based on data processed within the LIMS itself.

JUSTIFICATION FOR A LIMS

A LIMS is often justified on the basis of improved productivity. This concept needs to be examined carefully, since by some it may be interpreted as permitting a reduction in laboratory personnel. This is rarely the case, certainly not in the short or medium term. Indeed, when one takes into consideration the time that may be expended in planning for and implementing a LIMS, and the continuing need for support personnel, the system can hardly be considered a laboratory cost-saving device. In some cases, use of the system may actually increase the effort spent by individual analysts who must take the time to enter into a computer terminal the data that are obtained from nonautomated tests or instruments not connected to the LIMS.

What a LIMS can accomplish for the laboratory is allow for more efficient scheduling and use of resources. It also produces a more comprehensive and uniformly structured data base for access by management and clientele, provides systematic documentation, greater accuracy, and faster service. The net effect is that overall the laboratory is a more efficient organizational unit, not by virtue of more samples analyzed in less time—although this may in fact be the result—but by an improvement in the type, quality, and timeliness of the information it furnishes.

COST OF A LIMS

During the late 1970s and early 1980s, when the value of a LIMS first became widely recognized, the comparatively few systems that were implemented were developed in-house, usually evolving over a period of time as additional capabilities were recognized. In most cases, an enormous amount of work was required—ranging up to years of effort—and only relatively large laboratories had the incentive and technical resources to undertake the task. With the advent of commercial vendors the term "turnkey system" came into use, suggesting to some that a system could be purchased in much the same way as a plasma spectrometer or NMR unit and put into use with only a modest amount of additional effort.

Although it is possible that a laboratory's procedures can be adapted to a commercial system without modification, it is extremely unlikely. Despite the fact that certain operational features may be common to many laboratories, no two laboratories are sufficiently alike that the need for some customized functions will be avoided. The nature of the laboratory operation, the variety of samples and the way they are handled, the methods and instruments used, the number of work stations, the review, approval, and reporting mechanism, and the information to be maintained in archives are but a few of the factors that determine the degree to which a "turnkey" system will have to be tailored to meet the laboratory's needs. Nevertheless, because certain features common to the needs of all laboratories are now available in commercial systems, it is probable that few laboratories will opt for developing one entirely in-house.

Some suppliers' literature cite a price of $10,000 or less for a multiuser P-C based system. This is only a small part of the ultimate cost of installing and implementing a LIMS. What it does not consider is the customized features that may need to be added, either by the vendor or by laboratory personnel. It also fails to take into account the effort that must first be put into analyzing all facets of the laboratory operation and defining the objectives to be met by the installation. In a large multifunctional laboratory months may be required just to define the sample flow. The time value of those efforts alone may amount to many times the out-of-pocket cost, depending on the size and complexity of the laboratory operation. Despite this, in view of the contribution a LIMS makes to the overall efficiency of the organization, many laboratories are finding that the costs are manageable.

PLANNING FOR A LIMS

In defining the desired properties of the LIMS installation it is essential that the planning team consult with all potential users during the early stages of consideration. Clients as well as laboratory personnel at all levels should be included in the discussions. This will help ensure that all of the factors involved in the selection and design are considered; at the same time it will minimize possible resistance by those who may find it difficult to adjust to a change in operating procedures when the system is installed. These consultations should provide the planners with detailed information on topics to be discussed with the vendor.

The vendor's initial concern is likely to be with the major function of the laboratory. The requirements will naturally vary with the type of laboratory. Operations in a research facility, for example, will be more complex than those in a control laboratory, whereas in the latter speed of reporting may be more important. A technical service laboratory will have certain requirements different from either one. A laboratory engaged in regulatory testing will operate under constraints that are absent in other types of laboratories. Whether the customization is done entirely by the vendor, in-house, or both, the planners should be prepared with answers to certain questions such as those listed below.

Sample handling procedure. What is the pattern of the sample flow through the laboratory? Is all testing on a given sample done at a single work station or is it done sequentially at several?

Number of samples handled in a day, a week, or a month. Is the sample flow uniform or does it vary drastically? Is the sample load likely to increase beyond the present level and, if so, to what extent?

The maximum and average number of tests requested per sample. What are the different tests performed on those samples? Which are commonly performed as a group? How many are done routinely and is the number expected to remain constant for the foreseeable future?

Information obtained from the sample submitter. How are the samples described by the submitter? Are the samples delivered by the submitter or by someone else who should be identified? Are sample disposal instructions to be included? Is a job or project number included for chargeback or billing?

Identification of samples in the laboratory. What is the sample numbering system? Is bar coding to be used and, if so, will preprinted labels be used or should preparation be included in the LIMS? Will it be used only for logging in samples or will it be used for sample tracking as well?

Data entry. What data will be entered manually and what from instruments? What instruments are now in use and what is the likely expansion in the future?

Data presentation. Is quality control information needed to validate results for trend analysis, legal, or regulatory purposes and in what form is it to be presented?

Review and approval. What review and approval mechanism is to be incorporated and who will be responsible?

Networking. How many work stations or terminals are to be included? Are others outside the laboratory to have access to data and at what stage? What security needs to be provided to control access?

Client reports. How is the report to be formatted and transmitted? Is it to include cost information or is separate billing to be prepared? Are different forms needed? What information is to be reported in graphical form?

Management reports. What will management need to know about backlog status, productivity, and resource utilization? Are these on the basis of individuals or organizational units? Is client information to be referenced by job or project number, individual, user group, product tested, etc.?

Archiving. What information needs to be retained in the archives and for how long? What are the requirements on documentation for regulatory, patent, or other purposes?

A number of writers have discussed the factors involved in deciding to install a LIMS and implementing it.[2-5] The need for comprehensive planning and the time value of that effort have already been noted. The caliber of the personnel assigned to the implementation project is another consideration. There is general agreement on the necessity for a thorough analysis of the laboratory's operation, a clear definition of the short- and long-term objectives, and participation by competent personnel in the customization effort. Performance specifications and the responsibilities of the vendor vis-à-vis those of the laboratory staff should be defined in detail. Among questions to be addressed in the contract are

- Who will be responsible for interfacing instruments with the system?
- What modifications to the commercial package will be done and by whom?
- Who will be responsible for the initial training?

- What provisions are made for maintaining the system and responding to hardware or software failure?
- What is the vendor's commitment to providing long-term support?
- What are the provisions and timetable for acceptance testing?

Another factor that needs to be considered is the need for capable laboratory personnel to provide ongoing support. A system manager should be designated, particularly if a number of work stations are involved or if the system is networked to other operations. In any case, personnel should be available to train new users and handle at least minor problems.

It will be apparent that implementing a LIMS is not a simple task, and just because a system can be adopted doesn't mean that it should be. There should be a clear recognition of the additional effort that will be required for a smooth transition and to operate and maintain the system once it has been installed. Unless these added responsibilities are offset by compensating benefits, the results are likely to prove to be an expensive disappointment.

REFERENCES

1. **Dessy, R. E.**, *Anal. Chem.* **55**(1):70A–80A; (2):277A–303A (1983).
2. **Mizroch, M.**, Planning and evaluating a LIMS, *Am. Lab.* **20**(9):29–32 (1988).
3. **Mahaffey, R. R.**, Laboratory information management systems — pitfalls and justification, *Am. Lab.* **21**(9):40–44 (1989).
4. **Kobrin, R.**, *Scientific Computing & Automation*, June 7–8 (1989); Oct. 33–37 (1989).
5. **Megargle, R.**, Laboratory information management systems, *Anal. Chem.* **61**(9):612A–621A (1989).

Appendix

Table of Selected Net Present Values
(What $1 to be received in future is worth today)

End of	Interest Rate for Period (%)						
Period	6	8	10	12	15	18	20
1	0.9434	0.9259	0.9091	0.8929	0.8696	0.8475	0.8333
2	0.8900	0.8573	0.8264	0.7972	0.7561	0.7182	0.6944
3	0.8396	0.7938	0.7513	0.7118	0.6575	0.6086	0.5787
4	0.7921	0.7350	0.6830	0.6355	0.5718	0.5158	0.4823
5	0.7473	0.6806	0.6209	0.5674	0.4972	0.4371	0.4019
6	0.7050	0.6302	0.5645	0.5066	0.4323	0.3704	0.3349
7	0.6651	0.5835	0.5132	0.4523	0.3759	0.3139	0.2791
8	0.6274	0.5403	0.4665	0.4039	0.3269	0.2660	0.2326
9	0.5919	0.5002	0.4241	0.3606	0.2843	0.2255	0.1938
10	0.5584	0.4632	0.3855	0.3220	0.2472	0.1911	0.1615

Index*

---

* Page numbers in italics refer to case histories

Milton Keynes UK
Ingram Content Group UK Ltd.
UKHW040053071024
449327UK00019B/531